RAIN COMIN' DOWN

RAIN COMIN' DOWN

WATER, MEMORY AND IDENTITY
IN A CHANGED WORLD

ROBERT WILLIAM SANDFORD

EPCOR Chair, Water and Climate Security
United Nations University Institute
for Water, Environment & Health

RMB

For information on purchasing bulk quantities of this book, or to
obtain media excerpts or invite the author to speak at an event,
please visit rmbooks.com and select the "Contact" tab.

RMB | Rocky Mountain Books Ltd.
rmbooks.com
@rmbooks
facebook.com/rmbooks

Cataloguing data available from Library and Archives Canada
ISBN 9781771603171 (paperback)
ISBN 9781771603188 (electronic)

Printed and bound in Canada

We would like to also take this opportunity to acknowledge the traditional
territories upon which we live and work. In Calgary, Alberta, we acknowledge
the Niitsitapi (Blackfoot) and the people of the Treaty 7 region in Southern
Alberta, which includes the Siksika, the Piikuni, the Kainai, the Tsuut'ina and
the Stoney Nakoda First Nations, including Chiniki, Bearpaw, and Wesley First
Nations. The City of Calgary is also home to Métis Nation of Alberta, Region
III. In Victoria, British Columbia, we acknowledge the traditional territories of
the Lkwungen (Esquimalt, and Songhees), Malahat, Pacheedaht, Scia'new,
T'Sou-ke and W̱SÁNEĆ (Pauquachin, Tsartlip, Tsawout, Tseycum) peoples.

We acknowledge the financial support of the Government of Canada
through the Canada Book Fund and the Canada Council for the
Arts, and of the province of British Columbia through the British
Columbia Arts Council and the Book Publishing Tax Credit.

For Vi and Vi's Guys,
the informal local group of
wonderful musicians who have
become a reed through which they
sing themselves and the world
from grief into wholeness and joy,
and whose inspired company and
enthusiasm for playing together
motivated their number one fan to
endlessly hum the words to "Have
You Ever Seen the Rain?"
And to write this book.

CONTENTS

INVOCATION
Rain Comin' Down
9

ONE
Celestial Rivers
15

TWO
Rivers of Cold
47

THREE
Rivers of Heat
65

FOUR
Rivers of Words
93

FIVE
The Heart of Dryness
111

SIX
Irrigating Eden
137

SEVEN

Rivers of Memory

227

EIGHT

Rivers of Ice

249

NINE

As the World Burns

273

TEN

Learning from the Burning: The Summer of 2018

285

AFTERWORD

Rivers of Hope

311

APPENDIX

A Canadian National Glacier Act

319

BOOKSHELF

327

INVOCATION
RAIN COMIN' DOWN

I have spent of lot of time watching and thinking about water. This was not because I was predisposed to do so, but because of the importance of water to me personally in terms of how it gradually established my identity, how it played such a vital role in forming the country I live in, and how the presence or absence of it is now shaping the future of virtually every other nation on Earth. It could be said that this book is about how I caught up with how important water is to life and to what I was doing with my life.

If you are lucky enough to live where water is not generally scarce – if, say, you live in a country like Canada as I do – it would be easy to imagine that you have the luxury of taking abundant water for granted. We live in a changed world, however, where no abundance can ever be taken for granted, now or in the future. The surface of our planet and the composition of its atmosphere have been so altered and continue to change so quickly that even in Canada we are compelled to think differently about water, and to value it as we never have before. We are waking up to the fact that there has never been a time in the history of this country when it was more important to know where the water you drink comes from; how much of it we use and for what purposes; and the condition in which we return it to the river for others downstream to use. We are also waking up to the fact that in the future water is about to become more precious than we can imagine today.

My life work is to translate complex scientific research outcomes with respect to water and climate into language the average person can understand and that decision-makers can use to craft timely and durable policy. In this work, the most important thing I have learned is that if you want to understand climate change, follow what is happening to water. If our society is the ship we have to steer to avoid ending up on the climate rocks, then water is the medium through which we must slowly and carefully turn our runaway *Titanic*.

The reason why we follow water is because its remarkable chemistry and molecular properties shape the character of the world around us. We follow the water because in one form or another it is all-pervasive; seen or unseen, it is omnipresent. It is in the air as a vapour, on the ground as lakes and rivers, and *in* the ground as soil moisture and groundwater. It is present in or around every living thing. It is also in us and in all our food. Pull on any thread in the fluid fabric that holds our world together and you will see that thread connected somehow to every other one, and all of them connect in some way to water. This book is about these threads.

One cannot but be surprised, when studying water seriously, where that interest can take you. As this book will demonstrate, it takes you from the very origins of the cosmos right down to how the unique structure and remarkable qualities of water as a molecule shaped the formation of the Earth. You can know what it is like to ride Earthward on the tail of a comet. It takes you to the depths of the global ocean that came into existence because of comets, and by way of evaporation into the upper reaches of the Earth's atmosphere. An inquiry into the nature of water can carry you on the winds to the centres of storms, inviting you to imagine falling with raindrops to Earth and then gathering with other raindrops in tiny streams that flow into great rivers which sometimes go round and round in lakes. The study of water takes you into the very roots of trees, into the soil and into the dark courses of ancient underground rivers. It takes you from the thirst of one person to the thirst of nations, and from the demographics of the past to how the availability of water may alter patterns of human settlement in decades to come.

I have written that our entire Earthly reality is defined by all the ways in which water reacts with nearly every element in the physical world. Change even one of the critical parameters pertaining to water, and the world around you, no matter where you live, is suddenly a different place. One parameter, however, has more influence than almost all the others over the nature and behaviour of water. Change the temperature of water even slightly and an entire new geometry is created. The changing of a single defining parameter – temperature – alters all of the other biogeochemical properties that define the form water takes and how it acts. More and more we are seeing that this is exactly what is happening now.

It is also important to recognize that our Earthly reality is also increasingly defined by all the ways in which humanity reacts with and affects water. The study of water takes you to exotic places in Africa, Eurasia and the Middle East where water has been or is now an issue of national concern. Remarkably, an interest in water also takes you to a surprising number of places where water security will clearly be a critical issue in the future. Unfortunately, this means you cannot avoid politics. Understanding water means engaging in politics at all levels, whether municipal, regional, national or international. This is not something scientists are often comfortable doing.

Science has always avoided advocacy, for advocacy is often seen as self-interested and without objective basis. Scientists have instead allowed facts to stand on their own, in the hope that public understanding and government policy would eventually catch up with truth. At this moment in history, however, that doesn't appear to be happening. This is forcing scientists to re-evaluate their role in society in the face of a growing number of linked threats to the future of all of humanity. That said, scientists do not take any advocacy lightly. At the moment, this observer would characterize scientists not as advocates but as expert witnesses – truth-tellers who care profoundly about the world and who are willing to offer testimony based on the privilege of coming to know the world and the biodiversity-based processes that ultimately permit life to flourish there. These witnesses will no longer remain silent. Again, this is not an easy role for science to play.

Despite huge investment in research, there remains little urgency in dealing with the fundamental cause of changes in the behaviour of the global water cycle, namely carbon dioxide emissions related to energy production. We as a society are not taking the acceleration of the global hydrological cycle seriously enough. Part of the reason for that is that there are people who don't want society to take the threat of hydro-climatic change seriously. In the post-truth era in which we live, it has become customary to overvalue opinion, preference and ideology at the expense of proof and data. We are trending as a society toward decision-making anchored on a priori, nearly instinctive narratives – decision-making based on what is widely held or can be made so rather than on what is objectively true.

It is very interesting to observe how, in a post-truth world, society handles uncomfortable truths. One of the most widespread ways of coping with what we don't want to know is denial. The resistance in our society to acting on the climate threat is well financed. After all, there is a lot of money at stake, now and into the future. An entire industry has emerged, complete with internet trolls and botnets whose job it is to ensure doubt remains about whether or not humans are causing the global climate to change. And those voices are not going to give up anytime soon.

The post-truth mentality undermines 500 years of progress in science. These economic interests not only undermine the scientific method and the fundamental tenets of the Enlightenment but also threaten the rights of all to dignity and to healthy environments in which to live and raise children. Our current focus on short-term approaches that champion endless growth is in complete opposition to the evidence science has built up with respect to the nature of planetary existence and what the Earth System can tolerate before it collapses. Given the threats we face to the sustainability of civilization, there could hardly be a worse time for science and politics to become isolated by political leaders into separate solitudes. So what should one do?

It is time for scientists and citizens who stand by their truth to speak up. We need to accept our moral responsibility as a global community. The great value of science resides in its honesty and integrity and in its demonstrated capacity for world-scale co-operation. Our challenge is to demonstrate to society that humanity only exists by virtue of its being embedded in an interconnected Earth System in which water is a central connecting element. We face vast challenges posed by the decline of biodiversity-based Earth System function that can only be addressed through close global co-operation. Scientists can no longer remain silent. We must speak up. In this book we do.

But it must be pointed out that ours is not just a story of alarm. There is also stunning complexity and great beauty in what you can observe in following what water is and does in the world. And there is hope. Following water takes one back and forth in time, linking us to what the Earth was like in the past, what our reality is now, and how our understanding of water will shape our future. If we follow what is happening to our water, it will reaffirm the miracle that is the world. To fully appreciate the beauty of the Earth System and understand how water and climate are interconnected, we have to begin at the beginning – as this book does – with how Earth became a water world.

ONE

CELESTIAL RIVERS

For an extended period in my life I chased comets. It was an intellectual sport, a cosmic amusement that served as an excuse to travel. It took me a long time, however, to realize that in running after comets I was actually stalking water. I learned a lot of other things along the way. Comets were a way into mystery, into poetry, into deep science, but also into dark places in the human heart. But in the end, what pursuing comets left me with was an unshakable amazement – a sense of the miraculous, if you will – that compels me to continue to link what water is, does and means in the physical world to what is happening in that explosion of aspiration, competing symbols, needs, numbers and relativities that have begun to radiate hot and bright from the supernova that is human presence on this planet. There is much we can learn from the comet story that will help us put the meaning of this supernova into relief.

In chasing comets, you quickly come to understand that relativity is not as complicated as people think. Understanding relativity is made easy nightly. All you have to do is go outside and look up. The light that falls on you is so old that the stars which radiated it may have already died before the first humans walked on the face of this blue planet. We live in a galaxy so huge that it takes light more than four years to travel the 40 trillion kilometres that separate us from Alpha Centauri, our sun's nearest neighbour. From this we can understand the universe as an infinity of present tenses.

Water, in the form of comets, appears to have existed from the beginning. A flattened disc of colliding rock and ice surrounded a young star. The astrophysical politics of collision created broken rock and ice that swirled in their own orbits. Over time, comets sacrificed themselves to gravity, creating planets and punching craters into moons. Larger objects gradually coalesced and acquired even greater gravity, enough to allow thin gases to gather into planets of Jovian scale. Closer to the sun, warmer accretions captured thicker atmospheres, formed terrestrial globes. With the gradual consolidation of Big Bang debris, the number of worlds declined. Millions became thousands, then hundreds, and eventually, in this part of the universe at least, just eight (definitions of "planet" are nuanced in the context of bodies like Pluto). Water, it appears, found its way in one form or another to each of the outer planets. Comets with long orbital periods falling back toward the sun no longer recognized the place with its perfectly trimmed skies, immaculate cometary avenues – all circles within circles with much of the evidence of the origins of worlds now gone. And what was most astonishing is that one planet – the third one from the sun – was so perfectly situated in space as to be almost completely covered by liquid water. I went to Australia to see evidence of how this happened.

On a wall outside international customs, a sculpture, a shield circle – silver – with metal shards blown tailward. An airport Halley's. In its glittering wake are fingers pointing to the sky, truly an international arrival. Evidence of the presence of other comet-watchers is everywhere. There are ghost-star coffee mugs, key chains, pins and T-shirts celebrating Halley's

World Tour. A grand sky splash has drawn the upwardly mobile southward.

Sydney is wreathed in mist. A sunrise of Down Under proportion reveals a whole continent fated to look up at the sagging belly of a blue planet, an entire nation living at the base of a spinning celestial statute. Kangaroo fur ruffles in the wind created by the world's turning.

With my first look into the centre of the galaxy I know why we call it the Milky Way. It is a swath of stars so bright they cast a shadow in the absence of the moon. It is so bright, in fact, that Aborigines named the dark spaces reverse constellations. An elder now on her second time round Halley's orbit recalls being held up to the sky as a child in 1910 when the bright streak blistered a third of the Australian sky. Now in her mid-80s, she thinks her eyes are failing. She can barely see the comet – this time only a celestial fuzzball – but still claims it is the viewer who imbues every sky with meaning.

We stay up all night to watch the comet, and spend all day visiting wineries. I discover my own Vina Kometi. Drink gives one the courage of new gravity, arranges different orbits for ill-fated moons, brightens the fortunes of champagne stars. Who in their drunk mind could look up at that otherworldly streak and not see a cosmic cork released from a bubbling sidereal bottle. The comet is a veritable Bollinger of the heavens, a Veuve Clicquot of the predawn light with a comet burned in every cork. I think of the champagne we drink as the legacy of vintners' gratitude to the Great Comet of 1811, the spectacular sparkle of which was said to have ended a ten-year blight.

Life lessons are contemplated over wine. Meaningful events are separated in ways age puts into relief. Relative to the life of a comet, an 80-year-old woman has lived only a short time, a dozen tender hours that don't even make up a week of comet-ary days. But time does not order the relationships of mind. Instead the mind orders itself implicitly, connecting dusk's glow on Canadian mountains and kangaroo fur wimpling in the dry dawn wind. With the help of the wine we arrive at a summary cosmology. We float for eons as possibility only, then rise briefly, sleepwalking on the tightrope of time's arrow through sunlight space till our consciousness contracts inward toward a black hole through which we are absorbed in death. There is talk about inner and outer space. From the smallest particles to the largest objects in space, we marvel at the comparable emptiness: the distances that separate electrons from nucleus protons and neutrons; the same relative distances between stars and galaxies; the spaces of widening separation and isolation woven into the fabric of time, the warp and woof of our lives. We are hardly the first to be moved in the direction of such thoughts by the appearance of a comet.

I think of our long amnesia. Four million springtimes. We were too busy surviving to count. Children often died as soon as they were born. There were plagues. We fought. Meanwhile a hundred thousand comets came and went. In our restive nights we lived only under stars and a single, battered moon. Then, somehow, we became aware of the music of the spheres.

Think back to that late Pleistocene night when a Cro-Magnon first started noticing the night sky. Suddenly the universe spun; the leaves of trees and the ground beneath hummed. Rocking to this new beat, those who came long before us spent night after night pushing pins into the dark map until images of their own world emerged in the sky. The stars spawned enormous bulls, the bright belt of a heroic hunter, winged horses, fabu-lous dogs – a first cosmology snatched from the teeth and claws

18

of night sounds. We knew Pegasus before we knew the shores of our own seas. We yearned for the stars before we knew the oceans in our own souls.

The first true Knights of the Smoking Star held comets to be portents. While the rest of Asia slept, Chinese insomniacs stayed up every night for 1,100 years classifying comets into 29 categories by appearance and the mayhem they portended. Their taxonomy tells it all. A comet with four tails brought disease to the world; one with three a collapse of the state; two, a good crop of corn but maybe also a small war. This knowledge the court kept secret for fear that freelance astrologers or traitors in the Imperial Observatory might aid rebels and perhaps overthrow a king. The Chinese were the first to turn astrohistory into a basis of statecraft, but they were certainly not the last.

Aristotle argued the case for celestial immutability. Godlike, he saw the heavens as changeless. The world stood still in space with stars wheeling round it daily, the lower atmosphere clinging to peaks and tearing at trees while the upper ethers swirled with the stars. Heated by the sun, volcanic gases burst into flaming broom stars, comets, the auroras and the Milky Way. Though Aristotle's geocentric sense of the cosmos and simplification of the heavens overruled the logic of eyewitnesses, his views were held firmamently, religiously, for 1,500 years.

Not everyone agreed with Aristotle. Democritus stood Christlike among the superstition merchants of his time, sermonizing on the vibrating world of atoms and how their structural themes repeated themselves throughout the universe. The Milky Way was but a billion stars like our own, humankind only a roadside stop on some cosmic zoom. The world forgot Democritus for two thousand years before discovering that what he held was true. Then, within a period brief as his lifetime, he was elevated up the rocket gantries that put people on our airless moon, if only for a short visit.

In that long interim, the idea of Aristotle's unchanging sky was challenged briefly when Seneca dared to ask why the motion of comets was unaffected by the wind. The solar system winked with approval when Seneca went on to inquire whether comets might not come from without. Under cross-examination the universe began to assert its rightful dimensions. Unfortunately Seneca was also greatly interested in other heavenly bodies. The Emperor Caligula was not amused when he found Seneca on top of his sister when he was expected to be spending his nights in other rapt pursuits. Seneca was lucky Caligula did not cut off the offending body parts, as was certainly his custom. Seneca was instead banished but allowed to return to tutor Nero, which he did, teaching him all of the tricks of tyrannical intrigue. When a comet unexpectedly came it was taken to be an exclamation portending Nero's death. Rather than sticking to the science, Seneca allowed a self-seeking court astrologer named Balbillus to advise that monarchs could forestall the stellar omen with deaths other than their own. Nero, having murdered his mother, two wives and various other relatives, had experience in such matters. He ordered the Roman nobility to be wiped out. Under the ghost star Seneca followed the fiddler's mad command and took his own life for his burning state. He could not have guessed at the time that the apparition would pass, as well as Halley's six years later, until, under a completely clear sky, the evil king – the omen conductor – would descend to Hades by his own hand.

The craziness and the superstition, however, did not end with the death of Nero. Vespasian too was fearful of vengeful gods, and in response manipulated the interpretation of divine omens. As is often the case with official spin, Vespasian began to live fully within the kingdom of his own lies. When a flaming red star came in his time, he was surprised but shrewdly relied on what history now remembers as the divinity of the bald man. Vespasian contended that given his own baldness the hairy star came not for him but for the king of the Parthians, who had hair. And he got away with it. He also bequeathed this bald defence to subsequent generations with such superstitious success that even when medieval kings died natural deaths, peasants searched the night sky for the hairy tails of their Emperor's Stars.

We encounter the same superstitious beliefs about comets again later in the famous Bayeux Tapestry. Embroidered in it we see King Harold at the battle of Hastings. He is doubled over as if he knows that the celestial rocket in the heavens above was aimed at his rule. Though the fabric of this tapestry of doubt doesn't record what he was saying – it appears it could have been some futile order, some moot point. Clearly Harold was seeking an obvious sign that he could turn legends over and this comet had not come for him. Yes, but hadn't Jerusalem fallen at such an omen? Hadn't the mouths of Caesar's wounds cried out to such a star? Hadn't Constantine died under such night sky blaze, and Attila too, that Hun? Were not the Anglo-Saxons also of that sort? The year was 1066. Every schoolchild learns the story. Halley's reached perihelion during that dreaded March. And though the fateful battle was not fought until October, superstitions have no expiry date. William the Conqueror tossed the cosmic coin and Harold lost. Thenceforth a new star confirmed the arrival of a new king.

But then again there was Giotto, who dared to imply that comets could augur good. It is hard to know if he was being creatively faithful or simply looking centuries ahead when, in his major work of 1304, he painted haloed wise men around a manger. His *Adoration of the Magi* featured a star overhead, a terrible wound in a black sky, a blistered ball with a bloody tale: Halley's dressed up as a Christmas Star, conflated now with the Star of Bethlehem.

Giotto notwithstanding, superstition continued to rule. In 1456 Pope Callixtus III was deeply disturbed by yet another Halley's visit. Armies of Turks arrived in his dreams, bringing with them their One God. Each night was a horror of dark eyes and swords, terror he couldn't escape, for even as he awoke he was confronted by a waking dream of glowing dust and starlight still there in the morning sky. Callixtus excommunicated the comet, and though the Vatican denies this, he inserted it into the Ave Maria. Oh Lord, protect us from the devil, from the Turks, from the comet. (He could have added, from ourselves.) In August, with the evil overhead, he sent 40,000 against besieging Turks at Belgrade. Invoking exorcism, his Franciscans formed a wall of waving crosses and led the charge. The Turks withdrew, taking with them the smouldering anger of their defeated faith.

At the same time, burning stars were also troubling the Aztec sky. In splendid Tenochtitlan, Montezuma, fixated on legend, mulled the portended return of a white-bearded god. The stars said this god would want his kingdom back, though no one appears to have asked why he had readily abandoned it. Then two comets met in the sky. Though he did not yet know what such a creature was, Montezuma interpreted the burning midnight as a sign that he was riding someone else's horse. Each fire, each storm renewed the omen. The white god was on his way. The only ones surprised when the Spanish boats were carried onto the sand of the east coast were the conquerors themselves. Montezuma gave his kingdom back to a Latin Quetzalcoatl. Cortés and the Spanish, however, were not gods. In fact, they acted badly. Exhibiting barbaric greed, they enslaved the people and then melted the golden empire of the Western world into rings and religious figures, leaving intact only the myth of bearded gods arriving on the tails of comets.

The madness continued. Andreas Celichius, the Lutheran bishop of Altmark, preached in 1578 that "the thick smoke of human sins rising every day, every hour, full of stench and horror before the face of God, becomes gradually so thick as to form what we call a comet with curled and plaited stresses which at last are kindled by the hot and fiery anger of the Supreme Heavenly Judge."

The good bishop went on to claim he could smell brimstone in the devil's cometary tail. Andreas Dudith, a heathen apparently with a less developed nose, countered that he could not smell sulphur on the night wind and cheekily observed, not without logic, that if comets were caused by mortal sin, they would never leave the sky.

In the 16th century even the medical profession was eating the wafer and drinking the wine-red Kool-Aid of superstition. In 1528 Ambroise Paré, the great French surgeon, had looked to the heavens and witnessed a comet with a countenance so terrible that he claimed it made peasants fall sick in terror. The omen, the good doctor elucidated, was in the shape of a bent arm holding a great sword, its radiating axis formed of knives, bloody sabres, hideous faces, beards and bristling hair. Though there is no record of this, Dr. Paré must surely be counted among the fathers of psychiatry.

In the face of the Enlightenment, however, things could not

go on this way. In an Advent sermon Martin Luther affirmed the place of faith in matters of state and assured that his Reformation church would stay fearfully full by declaring – his pulpit quivering – that only heretics believed that comets arose from natural skies. God, he announced, does not create a single comet that does not portend a death. And yet, the year Luther died, a heretic was born whose views would empty the back pews and drive a stake right through the heart of religious superstition. Luther would have disapproved of Tycho Brahe, with his brass nose and the apparently psychic dwarf he held in his employ, not to mention his drinking parties that would go on for days – wild releases from Brahe's nocturnal stare into the eye of celestial truth. But in another sense, perhaps Luther would have forgiven Brahe had he seen that without so much as a telescope Brahe did for science what Luther was trying to do for Christianity: end the dogma in pursuit of ever-higher truth. No thanks to Luther, however, Europe continued to live under Aristotelian skies. But not for want of trying by Brahe. One dark night in 1572 a star brighter than Venus appeared in Cassiopeia, violating the papal decree that the map of the heavens must never change. Local peasants, their eyesight not impaired by higher learning, confirmed that a supernova, an exploding dying star, had appeared. It must have been difficult for the Church to deny the blinding impact the phenomenon had on faith. It shone for four months in broad daylight. Brahe placed the supernova not in the upper ether of the wind, but in a sphere among the fixed stars. The Church, too busy burning those who observed that the Earth was not the centre of all, ignored him, claiming he was crazy. But in 1577 a comet appeared whose orbit mad Brahe calculated to be beyond that of the moon. A few listened this time, observing that the days of divine kings might well be coming to an end.

There were other madmen. The manufacture of fashionable beaver hats in England and for export to Europe employed mercury, which with repeated exposure poisons the nervous system of workers, an occupational hazard so common that, as Lewis Carroll observed in *Alice's Adventures in Wonderland*, hatters were invited to tea parties just for the fine madness they brought. During that period, however, it wasn't just hatters that were thought mad. Isaac Newton evidently squandered his genius on alchemy, playing daily with lead, arsenic and antimony, and was said to be madder than any hatter. He also suffered paranoia, failed in normal relations with others and eventually was seen fit only for a job at the Royal Mint. But his was a fine madness. He was elected to Parliament, where in madness one would never be alone. Imagine the honourable member from Cambridge babbling the names of base metals; a tale about an apple falling; his invention of calculus, or perhaps more correctly his observation of the calculus of planetary motion and the rise and fall of tides; and his discovery that the orbit of the Earth around the sun is an ellipse. He told Edmund Halley that comets were divine consignments of fuel and water that kept the sun burning and the Earth from drying up. Newton's great book, the *Principia*, went unreviewed in his lifetime – no one understood it – but it remains today a supernova in thought still exploding into the future.

Seeing what is right before one's eyes sometimes demands forgetting the names of things seen. Jean-Dominique Cassini was the founding director of the Paris Observatory. Famous for discovering the gap in Saturn's rings and four of the great planet's moons, Cassini is also remembered for his hospitality in entertaining Halley and for his sharing of ideas. Cassini noted that three comets had arrived from the same part of the sky, moving at the same speed before rounding the sun in the same way. Cassini told Halley that these same three comets returned regularly. Others, who were not scientists, observed this also. Among these were the Bantu, sub-Saharan peoples who for a thousand years had claimed that comets came like visitors. They came, said hello, stayed a while, then left saying they would come back one day. Forget the omens. Forget the names. Watch the sky.

We arrive finally at Halley himself. A one-man scientific movement in his own right, Edmund Halley explored the age of Earth, founded geophysics, devised the first weather map and prefigured the idea of continental drift with his mapping of the planet's magnetic fields. Halley thought in colossal terms – he calculated the 150-million-kilometre distance to the sun – but was just as capable of thinking small as well. He attempted to measure the size of the atom; studied cuttlefish; considered ways of keeping flounder alive for midwinter meals; built the first diving bell and used it himself. While he was captain of a navy warship charged with making magnetic maps, he also learned to swear with great abandon and evidently so liked doing it that he swore all the time. He also took opium and then described its effects to horrified fellows of the Royal Society. Halley was also a close friend of Peter the Great, once getting so drunk that he pushed the czar of all the Russias through sleeping London streets in a wheelbarrow, taking out a few hedges while the pair laughed loudly into the outraged night. Meanwhile, great argument continued in Halley's learned circle about the rate at which gravity diminished as distance increased from the sun.

On a warm day one August, Halley decided to visit the cantankerous Isaac Newton in Cambridge. Newton blithely opened the conversation by offering that the planets orbit in ellipses and then showed the formulae by which he had arrived at this revelation. After applying Newton's formulae to the orbit of the Great Comet of 1680, Halley studied the paths of 24 other apparitions, including an unimpressive comet that he first observed in the early morning of November 22, 1682, when some said he should have been occupied otherwise with his new wife. Halley posited that this newly arrived ghost had been to Earth before, in 1607 and 1531. Halley projected that at aphelion – the point of its ellipse most distant from the sun – it ranged beyond Saturn, the most distant planet then known. Halley predicted it would return at the end of 1758, at which time posterity, he hoped, would record that it had been discovered by an Englishman. Halley continued on to other things, including teaching himself Arabic at the age of 49. He also devised the first actuarial tables and then violated what those tables said about life expectancy by undertaking, at the age of 65, an 18-year study of the cycle of the solar eclipse. Halley's consummate optimism was rewarded in that he finished the study a full two years before the cold January day in 1742 when, after one last hit on the bottle, he quietly died. It is ironic that the comet which astonishes the sky in his name every 76 years should be the ghost of perhaps his least achievement.

But the story does not end with Halley. While Halley was transmuting comets from omens into objects, Thomas Wright was getting himself expelled from school by imagining clockworks moving in Newtonian time. Wright's father believed that astronomy drove men mad, and he went so far as to burn his son's books. But neither his angry father nor the resulting lack of any support or formal education prevented Wright from later publishing *An Original Theory or New Hypothesis of the Universe* (1750). In this nearly forgotten work, Wright defined the nature and geometry of the Milky Way, not as a road of the gods, divine spilled milk or a pillar holding up the sky, but as a disc of stars floating in an ocean of space. He described the cometary nucleus as a corona of glowing gas surrounding and obscuring the great sweep of the tail as it wagged in the sun's magnetic wind. Wright might have been completely forgotten had his ideas not been pulled from the waters of obscurity by Immanuel Kant.

Kant too was swept up by Newton's innumerable stars and the dance of the singing heavens. He was 31 when his idea of an evolving universe dropped like a dead fish on the altars of religious immutability. There were not only countless stars, Kant said, but other galaxies, other worlds. Andromeda's swirling, Kant stated in 1755, was simply another Milky Way. He also offered that comets formed in the farthest reaches of space, and that their tails were composed of only the lightest of gases. When challenged with what vapours burned to form the comet's tail, Kant, obedient to the intellectual fashion of the day, riffled through the pages of the Bible. He found his answer in Genesis. Comets were water on the holy firmament. Perhaps overcome by such heady stuff, however, Kant made a terrible moral mistake: he forgot to whisper when he claimed the Earth moved and was not the centre. As a consequence, he could not avoid smashing headlong into the wall that was resisting Enlightenment. No, Immanuel, the Church admonished, you Kant say that. The pressure must have been unbearable, for in the end one of the greatest thinkers in all of history lost courage and reKanted his faith.

It is hard to fight the intellectual status quo. Superstitions are like the undead. They are difficult to kill, and even when you kill them they come back to life just when you least expect it. Daniel Defoe recorded that a faint and languid comet appeared just before a pestilence. Another one he reported, coloured a furious red and accompanied by the fearful rushing sound fire makes when burning dry wood, appeared just before the Great Fire. Defoe saw both stars as portents of God that reached down to scourge the streets of London. And, the presence of comets notwithstanding, that is what happened.

So where were women when all this was happening? It was 1758 and the projected return of Halley's was imminent. In France, astronomer and geophysicist Alexis Clairaut chose at the last minute to recalculate the comet's orbital tables, but he was running out of time. Clairaut engaged Jérôme Lalande to help. The work was actually done by Nicole-Reine Lepaute, however, who worked day and night for six months and concluded that the Halley's estimate was right. On Christmas night a self-taught German farmer with a homemade lens saw Halley's ghost inbound right on time. The comet became widely visible in March of 1759. The men took the credit, of course, claiming women should be valued for their beauty and their ability to oversee servants. Clairaut dismissed Lepaute's contribution by claiming she simply wasn't pretty enough – sexism on a cosmic scale, if you will. Lepaute just carried on, devising tables for parallactic angles, predicting an annular eclipse of the sun, applying the calculus of planetary motion to the changing heavens. Looking skyward through wide-set eyes, hoping for a glimpse of a distant age of reason. But sexism aside, perhaps such an age was not as distant as it might have seemed. Figuratively at least, it appears God was in trouble before Nietzsche killed him.

Pierre-Simon, the Marquis de Laplace, held that the solar system condensed out of the collapse of a rotating disc of gas and dust. He imagined the sun and the planets embedded in a cloud of comets, which would explain the amount of liquid water on Earth. He also surmised Jupiter's colossal influence in pulling the bearded stars sunward from outer space. Diffracted orbits, he called them. Hurtling snowballs grazed Jupiter's monstrous stormy face and were slingshotted at the sun, becoming captured forever between the two spheres. A flabbergasted Napoleon asked Laplace why his cosmic history possessed no God. Laplace replied that God had nothing to do with his hypothesis, and with this puff of breath dismissed a deity from the night skies.

Thousands of years of earlier accounts notwithstanding, the early 19th century marked the true birth of science fiction as a literary genre. At that time the salons of Paris had a howl at the expense of one Monsieur Montlivault, who proposed that the Earth was peopled from the sky. Montlivault had noticed the absence of human remains in strata rich in fossil life and surmised from this evidence that people grew up on the moon and rode comet tails to Earth. Landing on the sea, they sailed reed boats and learned to speak Egyptian at least well enough to order another round.

There remained a constant need to reconcile what humanity was seeing and learning with what was held to be the established status quo. In 1850 Thomas Dick brilliantly attempted to marry religious dogma to the discovery that humanity had new glasses and could see in the dark. Arguing that he saw clear evidence of the Creator's main design in the vast number of heavenly bodies, he surmised that their purpose was to provide habitations for myriad intelligent beings. Dick asked if comets shouldn't be seen differently. Should we not contemplate them with delight? Perhaps, he suggested, we should view them as splendid worlds of their own, vehicles – discount airlines if you will – conveying happy beings to new reaches of the divine realm.

Most enlightened Americans, however, were not buying attempts at the reconciliation of comets with religious humbug. Way out West a thoroughly modern George Custer was goading the Sioux to madness by invading their sacred grounds and disturbing graves in secret caves. Custer and his Seventh Cavalry regiment invited revenge. It would soon be over, they thought, this tiresome war with the Indians. Custer's men picked flowers, played ball, explored surrounding valleys. Shot whooping cranes. But suddenly a spectacular comet appeared the near the Big Dipper, a nebulous exclamation that startled the Dakota night. "A comet is a comet," Custer said, writing affectionate letters to his wife which began "My Sweet Rosebud." Custer's last stand.

By 1894 comets were glowing regularly in literary minds around the world. In one of his novels, George Griffith managed to dispose of the villainous Olga Romanoff with almost surgical precision by having her struck by a cometary bolt. In 1906 H.G. Wells invoked whimsical vapours from the tail of a visiting comet that stimulated an outbreak of love, reason and peace throughout the world. He was not the only one to presage what would happen when Halley's returned in 1910.

Comets orbited in Edward Barnard's dreams. They filled the skies of his sleep, but his alarm always seemed to go off before he could gather them. Then one night in 1881 the dream fulfilled itself for this amateur astronomer when from his backyard he saw a dozen tails, fragments of a comet crumbling as it grazed the sun. An eccentric American businessman, H.H. Warner, had offered a $200 prize for sightings of new comets, which Barnard claimed. He used the reward to take out a mortgage and build a house for himself, his wife and his mother. Coincidentally he found a new comet each time the note came due. Over the next 30 years he had found 16 comets and a fifth moon of Jupiter, and pioneered the use of sky cameras to capture his findings. On February 10, 1910, Barnard measured Halley's tail – it was five million miles long, he reported, telling the world there was reason to hope that by May 18 Halley's would reach to, and beyond, the Earth. The realization that the Earth would orbit through Halley's celestial foxtail unleashed a henhouse of hysteria.

William Huggins was an English scientist best known for his pioneering work in astronomical spectroscopy, which he conducted with his wife, Margaret Lindsay Huggins. The couple split the light from space into colours as a means of determining the composition and motion of the distant stars. After reading the spectrum of the burning arc of the Great Comet of 1881, Huggins pointed out the presence of carbon, hydrogen and the split bonds of exploding water in the bands of the spectrum. But he also saw something else: cyanogen glowing in its head. Huggins declared at the time to have reached the boundary of cometary knowledge, but he offered that the presence of a powerful poison in the head of the Great Comet did provide an opportunity for "enchanted speculation." Though Huggins saw value in his discoveries, few were interested in them until Halley's neared in the spring of 1910. A world still largely superstitious about comets pondered the prospect of passing through the comet's poisonous tail with a lot more than enchanted speculation. The tabloids of the day stoked widespread fear. Informed scientists told a panicked planet the Earth had passed through comet tails before. The Earth's upper atmosphere would dilute the small concentrations of cyanide to one part in a trillion. But Camille Flammarion – brother of the famous French publisher Ernest Flammarion – was an earlier Erich von Däniken, warning that all life would come to an end. But before that happened, the cyanide awash in the Earth's atmosphere would be converted to nitrous oxide – laughing gas. Hilarious, really, when you think about it: humanity snuffed out in anaesthetic death, a final cosmic giggle. Not everyone thought that was funny, however. Nations panicked. Millions took to the rooftops. Crazies began taking their own lives. Huggins missed the whole show: two weeks before the Earth spun into the graceful arc of Halley's tail, he died of old age with nary a trace of the offending gas ever detected over his fresh grave.

Still, the madness marched on, with the media leading the charge. Before there was such a thing as late-night news, the world got its stories first-hand, second-hand, third-hand or from rumours in the press. Countless newspapers vied for readers, and competing editors were not above making up stories on slow days. Engaging articles on topics such as eyeglasses for chickens were not uncommon. With the arrival of Halley's, articles about opposing scientific opinion concerning the nature of comet tails and the possible threat of the end of the world were written in the same tone as the social pages might report on who was present at an evening party at the home of a small-town mayor. Newspapers gave as much space to cranks and mystics as they did to scientists. Bold-faced lies made headlines. One newspaper invented a man named Henry

Heinman and his religious cult. Invented Jane Warfield, a pretty farmer's daughter, and then invented a scene in which the lovely Jane, naked and barely conscious, is tied to a stake. Invented a ring of frenzied dancers around her – Heinman's Horde. Invented a knife ready to be plunged into Jane's heart, a virgin sacrifice – an antidote to the comet's poison to prevent the end of the world. The newspaper also invented a posse and a brave Oklahoma sheriff who saved the poor girl. Yellow journalism of this kind, of course, could only have happened in 1910. Now we know comets do not have a scorpion's tail. Now the media does not invent facts.

With the urgings of the tabloids, the public took the prospect of *la fine del mondo* seriously. In 1910 there was nothing like the curse of a comet to push those already at the brink into insanity. Incessant brooding over the world's fate drove a man named Fred Bowers crazy. Another fellow, John Marlow, dug a cave with a shovel, built an airtight door and then herded his wife, his children, two horses, two dogs, a cat and his chickens into his gravel ark. All miraculously survived exposure to the comet's burning wake. A woman named Sophia House, however, turned on the gas to avoid celestial suffering. A melancholy Paul Hammerton had himself nailed to a cross – a crucifix against fear – but was saved from an Easter of insanity by his friends. Halley-lujah.

Naturally, the political cartoonists could not ignore the symbolism of Halley's visit. An Italian illustrator flattered Emperor Wilhelm II by depicting him wrapped in a greatcoat, white against a dark crowd, while around him ambiguous telescopes aimed at a glowing star. The Kaiser's Komet was depicted as a sky-tracing bullet, a cosmic cavalry charge, Germany rising above Europe's lesser star, the implication being that the world as it existed in 1910 was about to end. Another cartoonist depicted an upward-pointing czar who sees different constellations in a stormy Russian sky, the comet a burning fuse on a bomb of impending revolution.

Meanwhile, the comet was secretly imposing itself on the dim consciousness of Turkish street dogs. The Turks saw devils in eclipses and omens in every wind. The population of Constantinople gathered on rooftops each night, clapping to keep their children awake so they would not miss the last hours of the passage of Halley's tail and the end of the long-suffering world. In darkness below, police rounded up thousands of unlicensed dogs, and with the dawn came thankful survivors returning to suddenly quiet streets. To prevent Turkey from going to the dogs, an entire canine culture had been exiled, many of them made into gloves.

Many people simply drank their way through the crisis. Comet cocktails became wildly popular. Secret ingredients kindled a sensation of fire on the tongue and a warm feeling of floating in air. A jigger of applejack poured into a snifter of vermouth; a Nucleus Brandy; a Cyanogen Flip. Halley's Highball refills itself every 75 years. After a few of those even a blind man can see any number of blazing stars.

In 1910, Edward VII, Protector of the Realm, was King of Great Britain and Ireland, Emperor of India and reigning over a domain of 11 million square miles and 400 million subjects over which the sun never set. Many were thankful that Edward was bringing to an end the Victorian priggishness and public denial of sexual fact that went so far as to demand that tables have skirts lest there be any suggestion of legs. Going so far as to have past mistresses in the royal box for his coronation, Edward brought womanizing into the open. He went all out in everything he did: committed wholesale infidelities, shot more birds, ate more food, drank more wine, rode more horses and raced more yachts than any other king. Edward was indisposed, however, on the 4th of May, 1910, when Commander Peary, the disputed first man to the North Pole, was feted in London. The next day, Edward collapsed twice and could not resume his royal duties. Bronchitis got him two days later, but it could just as easily have been another ten-course lunch. Under the comet, British soldiers in Bermuda marked the king's passing with a 101-gun salute, at the precise finish of which the comet's head was said to have fired to a boiling red, marking, some claimed, the end of the king's short reign.

Famously, Mark Twain was also around in 1910. Samuel Langhorne Clemens had been born two weeks after Halley's 1835 perihelion, in Florida, Missouri, known to itself as the Show Me state. Show me, show me, show me, Sam demanded, and the world responded, revealing itself to the great writer in all its irony, contradiction, corruption and love. Twain became a mirror reflecting wisdom and laughter and was loved just as his characters were, across boundaries, languages and races. But life was cruel, taking three of his children and his beloved wife, Lily. And when Twain's talisman returned, it bore him away on its silent, remembered wind.

The year 1910 was a good year for comets. In January, South African railway workers discovered a huge apparition confused in the memories of witnesses with Halley's. It is forgotten now, though close to the sun it glowed in daylight. It remained distant, however. The Earth did not spin through its nine-million-mile tail.

It was as a child in the late 1950s that I took a direct interest in comets. While I was looking with excitement at comets in elementary school picture books, Fred Whipple was looking skyward from the Harvard College Observatory and saw what at first he imagined were glowing fishes darting through a cosmic sea. But as his resolution sharpened he began to see comets as swarms of gravel and sand in parallel orbits in space. Cometary heads were composed of sand coated with a volatile solid. When the volatile solid was spent, the fine sand fell in meteor showers, the iron burning red, the aluminum a hot white. Whipple glued the mass together. Sand and rocks embedded in ice – Whipple's "dirty snowball" – explained the splitting head and the capacity for long tails. Also the possibility of a thousand turns around the sun.

Only five months into the decade that brought the world the Beatles, environmental ethics and the alienation of youth, the newly born National Aeronautics and Space Administration was caught with its diapers down. Russia's premier, Nikita Khrushchev, announced to the world that a new kind of American plane had been shot down in Soviet airspace. To the folks at home in the u-s., it was spun as a great way to unveil the u-2. The State Department announced that this new plane, a flying experimental weather station, had been on a routine flight but had gone missing over Turkey. Khrushchev pointed out that the plane had been downed 2,000 miles from the nearest Turkish soil and introduced the name of the pilot: Francis Gary Powers. A weatherman evidently of the James Bond school, Powers carried a research lab comprised of a camera, a pistol with a silencer, enough Russian cash to buy a small house, three watches to trade and, according to Khrushchev, seven rings for comely Russian ladies. Also on his person was a poison capsule in case he got caught. The rest, as they say, is history.

While Apollo 13 made a distant pass just before its famous

malfunction, Comet Bennett was seen, fittingly, over Swiss glaciers. It was composed of a hydrogen head and a frail, glowing body.

The Egyptians, perhaps made myopic from too many troubled nights peering through gunsights 'and dodging exploding mortar shells, looked up at Comet Bennett and not surprisingly worried that what they saw might be a secret weapon ready to fall on them from the war-torn sky.

And the comets keep coming our way. I remember we expected much from Comet Kohoutek. It was a disappointing showing, but we learned a great deal from its visit. The average interstellar temperature is a few degrees above absolute zero, the temperature at which molecular action ceases, atoms collapse and your car won't start. In their long, slow orbits comets build up ice in the deep-freeze of space. Because of the cold the ices are exotic: carbon dioxide, heavy water, frozen rocks. In the comet's fall into the sun its speed increases and the sun warms the exotic ice. Passing Saturn, frozen methane vaporizes into gas. At Mars, water glows off into space. At Mercury, rocks boil in the cometary head. New comets are bigger and brighter than old ones. Fresh from deep space they are loaded with ice that has not yet met the sun. But with each passage a comet suffers a subtle death until there is little left to glow. Kohoutek's cold and distant light was a stellar disappointment to watchers on Earth but most especially to the Children of God, who predicted a collision and the end of the world to occur on January 31, 1974. But the comet turned out to be only a cheap celestial thrill.

Embarrassed by Kohoutek's poor performance and its failure to fulfill its headlines, the world media, which had hyped Kohoutek's arrival so enthusiastically, pretended not to notice the earthward-advancing Comet West, virtually banning mention of it in print or on TV. Spectacular outgassing preceded the breakup of its nucleus into four parts, which occurred in full nightly view of an oblivious Earth. Wags later mused that it was a portent of the end of the presidency of Richard Nixon.

According to Hoyle – not Edmond Hoyle, the English authority on card games, but Fred Hoyle, the British astronomer who formulated the theory of stellar nucleosynthesis – the modest influx of cometary debris in present times was but a feeble reminder of a planetary epoch during which comets filled the inner solar system with haloed light. Where did the waters of our global ocean come from? The in-fall of comets conferred the potential of life on our forming planet. These waters were soon swimming with life. Water that fell as rain also deposited cometary dust kilometres deep on the surface of our planet's crust. While deep within the Earth carbon hardened into diamonds, on the surface cometary debris crumbled into lichens, trees, flowers and finally blood. So far, so good. But then Hoyle, an expert in the synthesis of chemical elements in the stars, went a little too far. Working with Sri Lankan astrophysicist Chandra Wickramasinghe, Hoyle posited that the first life on Earth began in space and spread through the universe via panspermia, and that evolution on Earth is influenced by a steady influx of bacterial and viral spores that arrived via meteors and motes of cometary dust. Their claim that comets harboured life, however, was viewed as a protobiology of the fantastic and was widely opposed, though mostly by the mute cold of interstellar space.

But such debates did not silence the technicolor roar. Comets are small and therefore have little gravity. Outgassing of the comet's ice head as it approaches the sun creates a coma – a thin, luminous fog that can extend tens of thousands of kilometres behind the nucleus. As it moves closer to the sun, the halo of fluorescing hydrogen that is the cometary coma is transformed into an aurora of ions glowing an ethereal blue. Gravity pulls the comet toward the sun at 20 kilometres a second. The radiation from the sun creates an opposing hurricane headwind that at 400 kilometres a second blasts the tail leeward. The solar wind collides so hard with the thin air of the cometary globe that it creates a wave that thunders for a million kilometres in the wake of the comet. The comet becomes a cosmic jet breaking a soundless barrier. The sun's magnetic storms make the giddy tails behave wildly. They flap through space like weather vanes under a hot and windy sky. Nightward blown, the dust in the tail glows dimly yellow. Water vapour glows faintly red. During total eclipses, comets sometimes appear in the solar corona. Comets previously unknown – sungrazers, hitherto invisible – are bleached out by the glare. After an eclipse they are gone. A comet can also collide with the sun, and when that happens it is vaporized in a staccato of flashes, its tail waving headless in the nuclear heat, a snowball burning in hell.

Pursuing comets widely as I was then, I remember one night, while waiting for the stars to come out, contemplating the zodiacal light. I recalled that Seneca, before his unfortunate little affair with Caligula's sister, had called the sun's dusk afterglow a reflection of distant flames upon the sky. The fires of hell, perhaps. I am made speechless suddenly by the thought that the faint light through which the Earth spins along its ecliptic is generated by the constellations of the zodiac, a flat ring of stars that, Saturn-like, envelops the sun. What we see after the sun sets but before it's dark is in part zodiacal light illuminating a necklace of dust composed of the spent tails of forgotten comets, still glowing even in a faint and distant solar wind.

While I continued figuring out how to get to the right hemi-sphere to view the next comet, Jan Oort was engaged in some cloudy thinking. Using a radio telescope, he mapped the spiral structure of the Milky Way and established the Earth's relation-ship to one of its distant arms. He then proposed the existence of an immense and distant cloud of comets – ten million miles across and extending to the outer reaches of the sun's gravity – which he held was a remnant of the Big Bang that created the universe. Oort then demonstrated how a wandering star might create a gravitational flurry in the cometary cloud that would send comets careening down into the planets. He claimed the cloud was heavily populated with a trillion or more cometary nuclei – more, in fact, than there were stars in the Milky Way, but still an insignificant number given the immensity of space. Oort postulated that once in a geological age a big star might move through the inner cloud, shaking it like a fruit tree, re-sulting in a billion comets fluttering to Earth, one an hour for a million years. Rain coming down, indeed. This, of course, given the nature of scientific inquiry, raised the question of whether or not the sun might have a solar sister. Most stars, it appears, have gravitational sisters, one bright, the other a dwarf. Though far separated, their binary fandango can sometimes bring them together. If our sun had a sister – a dim red Nemesis that came to visit every 30 million years or so – her arrival would cre-ate a shower of millions of comets, a torrential rain from the Oort cloud into the orbit of the Earth, a spectacular present from a long-lost aunt, with each visit a great dying followed by a great birth. But our sun does not speak of lost relatives or unexpected reunions. Our sun does not point a familial finger at a sister in the surrounding stellar crowd.

It is now estimated that a third of the Earth's recently formed craters were created through collision with long-period comets, visitors from the distant cosmos that go bump in the night. The others are wounds rent in the Earth's hide by asteroids, debris from broken moons, boulders whose orbits the Earth interrupts in its paths around the sun. A small stone only ten kilometres across – a mere pebble compared to Goliath Earth – struck us, propelling iridium clouds and pulverized ocean floor into the sky in a darkening autumn that brought the Cretaceous to a close. On land, thunder lizards, dim-brained, their armour useless, wandered a frozen waste and then died. But surprise: unexpected possibilities in genes realized new forms that would inherit a world which, were it not for comets, would still be ruled by reptile kings.

Maybe it is the eclectic interests of the people who live there – their gregarious style, their money, their constant hype – but undeniably bizarre things in happen in California. Perhaps it is not surprising, then, that in 1977 Charles Kowal looked into the night from an observatory in that overstated state and discovered a cometary corpse beyond the orbit of Saturn. Chiron, a dark comet 300 kilometres across, its exotic ices all blown away, hid comatose, perhaps the last of a race of dying giants with which Hollywood could do a great deal. Filmmakers imagined the collision – the ultimate special effect. You would be able to read the paper by the comet's light, bold headlines reporting the last night of the Earth. But what is the real risk of cometary collision? Space technology, separate from missiles and weapons defence, has made it possible to create an accurate census of nearby comets and approaching asteroid belts that could impact the Earth. Means have been devised to deflect objects whose orbits might intersect with our own. Intruders can be nudged out into space to avoid dinosaurian implications for all life on Earth. This risk of a collision is small, however – one annual chance in a million, about the same risk we take whenever we board a plane.

With Halley's fast approaching for the 30th recorded time, the global scientific community determined what research would be done. Five spacecraft rose from the Earth, the work of 20 nations. One approached the day side of the comet so close that impact seemed possible. Imagine the Giotto space probe smashing into the comet's head at a relative speed of 150,000 miles an hour, glittering fragments embedded in the cometary snows and carried out past the orbit of Neptune like, as Ann Druyan brilliantly described it, Ahab's corpse pinioned to a cold white whale.

Instead of participating in the Giotto project, the Americans built a hundred bombers, the cost of any one of which could have launched a comet probe. But, to be fair, the Americans had already been to the moon. It is also true that at the time of Halley's visit in 1986, a modified U-2 plane was being used to capture cometary debris from the upper atmosphere – or so we are told.

Whether Halley's had anything to do with it or not, the visiting comet did see new tyrants deposed. In the Philippines, Ferdinand Marcos murdered his presidential opponent, only to find himself facing the sudden popularity of the victim's wife. Following an election which Marcos tried to fix, the victorious widow exposed his colossal graft as well as torture chambers and the mass graves of political dissidents. In Haiti, Baby Doc Duvalier, son of the first president for life, ruled in decadent splendour the poorest nation in the Western world. His rule was sustained by terror and police arriving in the night. But evidently some kind of voodoo caught up with the dull boy who suddenly found himself king of a country defined by his father's insatiable greed. Homeless on the global ocean of their own crimes, Marcos and Baby Doc searched every coast for even a small haven, but under the comet all lights dimmed with the stench of their approach.

America's Apollo missions taught the world much. Half of the mass of the Milky Way, for example, appears to be missing. The gravity is present – the influence of its mass – but when we look up, there is nothing there. No stars. No gas. No dust. Only the ghost of a missing something.

But despite unimaginable breakthroughs in science and new hope for the future, the post-Apollo era was a crazy age. It seemed nothing was impossible, no idea too far out there in space or time. The physicist and futurist Freeman Dyson looked to the skies and imagined millions of comets, abodes of water, carbon, nitrogen: constituents of life. He speculated that if the trajectories of comets could be altered toward the sun, trees could be made to grow on their sun-warmed sides. Bioengineered for the absence of gravity, well-watered trees in such bright sunlight could grow a thousand kilometres into space. In their leaves Dyson saw a potential return to our beginning: imagined countless arboreal earths, seeds sailing on the solar wind.

The post-Apollo era was interesting also in that humanity seemed to be drawing inward, our gaze diverted from the sidereal back to the terrestrial. Suddenly it was all about us again. I remember that after Halley's, other interests beckoned. Under the press and exigencies of life, an old flame died. In time I barely remembered glowing ecstatic when love's sun awakened a meteoric desire to penetrate the orbit of her blue world. But in my nightly confessions she saw only vaporous dreams. Years later I realized it was the sun I desired, and that she was held in place by the gravity of unseen expectation. In shopping centres and restaurants we sometimes pass, a fading comet passing a darkened Earth. But coming back down to Earth is not always a bad thing. Nearly ten years on after Halley's visit. I chased comets no more. But then one came to right to me. I remember it fondly. Urged on by the god of the open road, I manoeuvred again into the passing lane, the prize a fast time to a pointless city limit. "Shouldn't you slow down?" my wife asked. But before I could offer the witty comeback that was forming on my tongue, I saw a milky vapour reflected in the gleam of the hood. An expletive arose out of the ready wit. Then I saw it in the rearview mirror. In oblique afternoon light a crimson nucleus trailing a glowing contrail of steam. It was Hale-Bopp, my station wagon comet bound for a distant family star. Though I have yet to learn to stay home, I learned from that comet that much of what you need to know about water, and the world, can be affirmed right where you live. I have also learned since then that our focus should be on Earth now, on our home, for there is much that the water comets brought to this planet that is now telling us about the changing nature of the global hydrologic cycle. And it needs our immediate attention if we are to have a sustainable and prosperous future.

The lessons of the comet story continue repeating themselves. In our post-truth era it has become customary again to overvalue opinion and personal preference at the expense of proof and data. There is much the silently burning ice can tell us. The next chapters will make the case that there is no time in our history when it has been more important to abandon the myth, magic and popular opinion that obscured our understanding of comets for centuries, and take up the evidence generated by the careful methods of science. There is much we can do, but our future depends on our doing it.

TWO

RIVERS OF COLD

Ad in the *Rocky Mountain Outlook*, Canmore, Alberta, April 22, 2010:

Job opportunity: snow farmer

The snow-farming supervisor must be familiar with all ski area terrain with and without snow. This includes familiarity with all micro-features and all details pertaining to elevation, aspect, slope, angle and exposure to wind direction. They also must have an historical knowledge of weather conditions. Snow-farming work is micro-terrain-specific, and a difference of a few metres can have a huge effect. The snow-farming supervisor requires the ability to interpret weather and wind forecasts and use the information to create an effective snow-farming program, with the goals of the resort and run openings in mind at all times. Some of the requirements – read, responsibilities – of the position include but are not limited to: supervising a team of up to 20 staff and volunteers; supervision of the planning, setup and maintenance of 15 kilometres of snow fence; and competence in the basic operation of snow-grooming equipment. The successful candidate will have an intimate knowledge of the prevailing weather conditions and wind patterns, will be an expert skier and have the ability to speak proficient English. Must be dependable, mature and hardworking. Must also complete other tasks and special projects as assigned. Remuneration: $15 – and three cents – per hour.

It is not always rain coming down. Water takes many forms. By April many Canadians are weary of shovelling snow. Having done so since November, we find the winter has gotten to be very long. But just as I was about to throw my hands up in despair and defeat and join just about everyone else in town in complaining about the relentlessness of the Canadian winter, it struck me that I got more out of shovelling snow than I thought. I had to remind myself that some of the pleasure I enjoyed from those hours arose from actually seeing immediate and obvious results from my efforts, which happens rarely in my day job. I realized, however, that though I sometimes grumbled about the chore, the lift I got from those labours derived also from testing what I know theoretically about water against what the actual snow in my driveway was teaching me about hydro-climatic processes occurring right in my own front yard. This too was gaining knowledge, and it fed into my enthusiasm for doing direct observation rather than just reading about how the Earth's orbit, in tandem with evaporation, creates the weather patterns we all take for granted, or how the strength and incidence of sunlight on water in its various forms define the seasons.

So I gave some thought to writing a set of instructions for optimizing the pleasure of shovelling snow off driveways like mine. Something I could pass along to subsequent owners certainly, but perhaps also to neighbours who may weary of this simple chore, or dismiss it altogether as something they may wish to avoid by moving to warmer climes. It occurred to me that for those of this latter persuasion, such a set of guidelines could at least provide some indication of the pleasures and intellectual stimulation they were missing while they lingered day after day on some tiresome beach in Mexico. But then, in my mind's eye, I could see them each evening no doubt enjoying sundowners while shaking their gloating heads at weather reports from their summer homes in the Great White North. I could hear them far in the distance laughing maniacally at texts from frostbitten children and friends whom they had in their departure condemned to ride out the winter alone. I could hear them snickering, congratulating themselves on being smarter than those they left behind. My instructions, I decided, should be for those who stay, who are left behind in the dark, in the cold and snow. Often, those who overwinter thrive by turning it into a game – hockey, perhaps; skiing, certainly – but always there is one tedious thing we have to do, and that is shovel snow. One way to make that chore tolerable is to turn it too into a game, a game that like solitaire you only need one to play. You too can create your own shovelling game.

First of all, if you want any chance of winning the game, it is important to take snow shovelling seriously before you have to start doing it. If you don't, it can easily get away on you before you know it, which will almost assuredly make it a hateful chore long before the end of winter. You need to have high standards for playing this game, right from the outset of winter, or the snow will first demoralize and then defeat you. While others, including members of your own immediate family, may mock you for this, it is important to start the winter by establishing clear lines demarcating the area you want to keep clear for the duration of the winter. These sharp lines are not only practical but over time become highly aesthetic territorial expressions of one's will not to utterly give in to the darkness, the cold and especially the snows of winter. It is important to realize from the outset in late fall that these lines will be absolutely impossible to defend over the coming five to six months. Think of them as ideals like the UN Charter, your national constitution or the National Parks Act: high aspirations to which one must commit but which in reality are beyond the capacity of any one person or any human organization to ever fully realize. Expect that over the winter the sharp lines to which you have committed may become ragged and even obliterated. This will happen because you will not always be there when the snow falls, and the shovelling will fall to those with lesser standards or perhaps no standards at all. Accept that ground will be lost to those who are not as particular. Accept that over the early and middle arc of winter, the snows will always win. This does not mean, however, that strategy is unimportant, for it is strategy that will determine first your choice of snow shovel, which will in turn largely define whether shovelling is an irritating and inescapable drudgery or a pleasurable exercise for body and mind. Another strategy for not letting snow shovelling get away on you and become a hateful responsibility is to shovel early and often rather than letting snowfalls build up into an overwhelmingly ungratifying challenge of simply getting enough of it out of the way to survive. It is easier to shovel 5 centimetres of snow as it continues to fall than to wait until 20 centimetres has accumulated, making its removal so exhausting that the whole project appears meaningless, if not utterly unnecessary, in view of the options of where and how one can live in an increasingly globalized and far more mobile world.

I have found I can shovel our driveway, which is sizable but nothing compared to what exists elsewhere in my town, in eight to ten minutes if the snowfall is 5 centimetres or less. With this method I can deal with a heavy snow of as much as 40 centimetres in 40 minutes of short bursts of easy shovelling, whereas it would take more than an hour and a half at least to shovel the same amount of snowfall all in one go, if indeed I could even do that without a break. This has greatly influenced my choice of snow-clearing technology. Whereas I used to be committed out of little more than upbringing and habit to the single-handled snow shovel, I now favour what resembles a wider hand-held snowplow that can be pushed in long swaths to easily remove snow as long as it is not so deep or heavy that I am not strong enough to easily do so. What I have also found desirable about this method is that it allows for a longer uninterrupted push, which, along with clearing more of the driveway with each swath, adds to the pleasure of having clear, visible evidence of results.

Another key to strategic snow shovelling over the course of a long winter is to prevent ice from building up when, for example, a vehicle's weight compresses the snow under its tires into hard ice. While it is interesting that snow will do that itself under its own accumulating weight over generations and in so doing create a glacier, you don't want a glacier, even if only a temporary one, to form in your driveway. Even on a micro-scale like that, ice formed from cold compression can create jarring obstacle to smooth shovelling. A similar irritation can also result when clumps of hardened snow are knocked or fall from the wheel wells of cars and then freeze into cement-hard mounds that cannot be shovelled away or even penetrated by the sharpest metal edge of even the best snow shovel.

A little snow physics might be helpful to understand why the presence of hard ice clumps can confound even the best snow shovelling strategy. It all has to do with the structure of ice as it forms, which has otherworldly qualities not shared by water in its liquid form. As I wrote in *Cold Matters: The State and Fate of Canada's Fresh Water*, the presence of vast seasonal tracts of snow and ice has a huge effect on the global climate system. This is instantly obvious when you step outside in a normal winter almost anywhere in Canada. The first effect, which one immediately experiences, especially on a sunny winter day, is the extraordinary extent to which snow and ice reflect sunlight back into the atmosphere. This high degree of reflectivity – or albedo, as it is called – operates as a climate feedback system in and of itself. The expansion of snow and ice cover increases the albedo, thereby increasing reflected solar radiation, which lowers the temperature and thus enables snow and ice to expand even farther in extent. This is how winter gets a hold and keeps a hold until the longer days and warmer temperatures break that grip in the spring. This is important to know when you step out the door to commence shovelling. The force you are about to confront out there in the driveway is colossal. Ten trillion tons of water in the form of snow are transferred between the northern and southern hemispheres over the course of each annual seasonal cycle. Be glad all you have to shovel is the driveway.

On average, 23 per cent of the globe is covered with snow at any given time. While this is beginning to change with the warming of the global climate, what is not likely to change soon is the fact that the effect of snow cover is most pronounced where the land mass is concentrated the way it is in Canada. While you may not wish to bring this up with your unhappy neighbours who are out there – under protest – shovelling not for the pleasure of it as you are, but because they can't get their car out of the garage, you know the deeper reasons for why you are there. You are out there with them because the very presence of the snow you are shovelling has reduced mean monthly temperatures by as much as 5°C. The cold has brought the snow, and the snow has brought more cold. Nor is it likely to improve the disposition of your neighbours if you share with them the fact that over the past decades this effect has been working in the opposite direction. Even though they may be among a majority where you live who don't believe in human-induced climate change, they are likely to get the wrong impression if you tell them that snow and ice cover – despite wild variability – is shrinking, reducing albedo and increasing absorption of solar radiation, which is in turn increasing winter temperatures, which in turn will in time reduce snow and ice cover. In the Canadian Rockies, deep winter cold is less common. As one climate scientist put it, −30° is the new −40°. In the darkness, cold and knee-deep snow, neighbours can be unpredictable. You don't want to give them information that further frustrates them or gives them false hope that winter as we have come to know it may soon be a thing of the past. Prolonged exposure to cold, dark and snow can make people irrational. It is not out of the question that your frostbitten neighbours, misinterpreting the climate signal, could raise their shovels like placards and march together down the street in happy solidarity, never deviating again from the vision of warmer winters, but in so doing failing to understand that warmer winters would be only part of a larger and far unhappier future through which all of us would have to do way more than just shovel.

One of the ways in which snow and ice influence the global hydrological cycle is through storage of water in snow, glaciers, ice caps and ice sheets and through associated delays in the release of runoff from these sources. The time scales related to such releases range from weeks, as in the case of snow in and around your driveway, to months for high-altitude snow cover in mountains, to decades and centuries in glaciers, and to hundreds and even thousands of years when it comes to ice sheets and permafrost.

Snow is a reservoir that stores water for gradual release downstream in the spring in precisely the same way a dam works. This simple fact is of fundamental importance to a great number of people on Earth who without this runoff could not grow enough food to live. While you are really just moving the snow around, you should take pleasure in the fact that you are engaged with the substance of the global economy. There is not enough money in the world to build all the dams that would be required to store all the water the winter snowpack holds for later release into streams and rivers. Cold provides this valuable service for free. Snow is the currency of that free service. And in piling snow you have shovelled off your driveway into mounds which you will notice can last the entire winter, you provide this same service – albeit unwittingly.

If you watch carefully through a full winter, you will also notice you can tell a great deal from the size and consistency of snowflakes as they fall. The colder the temperatures, the smaller the flakes. Again, this goes back to the relationship between temperature and the amount of water vapour the atmosphere can hold and transport. Very cold air can hold very little liquid water, so the flakes are limited in how big they can grow. Larger snowflakes indicate warmer weather. Because the warmer atmosphere is carrying more water vapour, the flakes can grow bigger and bigger as they fall. Because these bigger flakes contain a lot of moisture, they can cause serious damage when they accumulate. The weight of what can be called "damp snow" can cause branches of trees to break and the roofs of buildings to collapse. Accumulated wet snow can also freeze into ice at night, which can cause considerable damage to roofs and buildings.

Fog and its children dew and hoarfrost are especially interesting. Again, cold air can't hold nearly as much water vapour as warm air can. Fog appears when air can no longer absorb water vapour, because it is already saturated. When temperature drops at night, already saturated air can no longer hold the same amount of vapour, so it sweats it out, so to speak, in the form of dew and fog if it is not too cold, and as hoarfrost when it is colder. This is why fog is more common during late fall, winter and early spring than it is in summer. The warm summer sun acts like a hair dryer. When the air around your wet hair is warmed it can hold more moisture, and when that happens, your hair dries more quickly.

Then there is the entirely other, and otherworldly, problem of ice. If you have read Kurt Vonnegut's novel *Cat's Cradle*, you will already be aware of how tricky ice can be. Vonnegut's brother, Bernard, was a researcher with General Electric, the huge and secretive conglomerate based in Schenectady, New York. Vonnegut himself also worked there briefly, interviewing scientists and writing press releases, and he used his experience there to create its metaphorical surrogate – the research laboratory of the General Forge and Foundry Company – which figures prominently in the novel. According to Cynthia Barnett in her book *Rain: A Natural and Cultural History*, Bernard Vonnegut was a cloud physicist who worked on a team studying why planes often lost radio contact while flying through electrical disturbances in snow storms. Almost accidentally the research team discovered that silver iodide dropped from an airplane could be used to precipitate heavy snowfall, which be-came the foundation for the science of cloud-seeding. In sub-sequent research, however, GE found that the results were often unpredictable and even dangerous, so it released all its weather modification patents to free itself from liability in the event that these technologies might fall into the wrong hands and be used to weaponize weather. It appears also that Bernard Vonnegut described to his younger brother the various states in which ice could exist, for "ice-nine" became the metaphor for the threat of nuclear war that relentlessly drives the masterfully intricate plot of *Cat's Cradle*.

The story revolves around Dr. Felix Hoenikker, a Nobel laureate credited with being one of the fathers of the atomic bomb. Upon being pressed by a military general to solve the 200-year-old problem of the U.S. Marines having to wallow constantly in mud, the intellectually playful Dr. Hoenikker sug-gests there might be a single microscopic grain of something that could be made to turn infinite expanses of mud instantly solid. Dr. Hoenikker ponders the problem, and after examining the several ways in which cannonballs might be stacked on a courthouse lawn, imagines the multiple ways in which certain liquids freeze and the different ways their atoms can stack and lock into a rigid, orderly structure in freezing. Dr. Hoenikker then observes that a tiny grain of a certain crystalline pattern can induce atoms to stack and lock and to crystallize and freeze in a novel way. He begins experimenting and finds there are many ways in which water in particular can be made to crys-tallize and freeze. The breakthrough comes when he surmises that water on Earth always freezes as ice-one because it is never exposed to a seed that would teach it to form ice-two, ice-three, ice-four and perhaps a form that might be called ice-nine, which freezes hard as rock and instead of melting at 0°C, melts at a temperature even higher than that at which ice-one boils. After this invention, however, Dr. Hoenikker discovers that if the ice-nine crystal is released into the world it will freeze every drop of water on Earth, destroying all life and bringing civilization to an end. The ending is breathtaking. *Cats Cradle* remains just as relevant today, as nations hostile to one another are now spending trillions upgrading obsolete nuclear arsenals just as Vonnegut – and the rest of the world – so rightly and desperately feared when he published his classic novel in 1963. Fortunately ice-nine does not exist – at least so far as we know.

This takes us back to the snow and ice in your driveway and how you don't want big clumps of the latter to take up residence there. So, let's say your friend Mike, who happens to be a glaciologist, visits and stays overnight. He has travelled through a heavy snowfall. He parks his snow-laden pickup truck in your immaculately shovelled driveway, but before he leaves after a hilarious night of drinking and outrageous reminiscences of what happens to researchers' behaviour after prolonged periods in high-altitude camps on various icefields in the Rockies and the Arctic, he kicks a clump of compacted snow from each of the wheel wells of his truck onto the driveway, and then with a hearty "hi-yo Silver, away!" he disappears into the storm that has followed him across the West. Unbeknownst to you, he has done this purposely to leave you with a lot to think about after his visit. What he leaves in the driveway in the midst of the heavy snowfall are effectively equations which he challenges you – the shoveller of snow – to solve in order to give greater meaning and value to your efforts to keep enjoying the sport for which you now have season tickets.

Against all the odds stacked against you by the laws of

physics, it is possible to beat the snow and ice at least sometimes. One of the most powerful influences in the entire global hydrological cycle relates to the amount of heat involved in changing the state of water back and forth from solid to liquid. Snow cover acts as a heat bank which stores and releases energy over time. We are so focused on staying warm while we are out shovelling that we can miss the fact that the global energy balance is about how the planet distributes energy, not about keeping us comfortable. You don't need to stand out there in the driveway very long during a blizzard to know that Earth System function is not about us humans. Irrespective of our presence and our influence, snow remains stubborn. You can't make snow do anything it doesn't want to do until it is ready. It takes a great deal of heat, for example, to melt it. The amount of energy required to melt 1 kilogram of snow that is already at 0°C is equivalent to the amount of energy required to raise the temperature of liquid water to 79°C. What this suggests is that if you get into an energy deficit with snow and ice, you are involved in global energy economics big time. Let clumps of compacted snow freeze in your driveway in December and you will have months to contemplate that economy. Barring a prolonged chinook, these clumps will stand in the way of smooth shovelling until enough heat can be poured into them by the April sun to melt them or until you take an ice pick to them. Withdraw the energy from a snowbank and you have to deposit it all back again in the spring if the account is to balance. If you keep this in mind as part of your shovelling strategy, nature will take care of the energy withdrawals and deposits for you without your having to expend a great deal more energy chipping away at clumps of ice in the driveway.

What one can learn about snow is cumulative, over both a

single winter and many winters. If you pay attention, you learn the difference between wet and dry snow and when to predict which. Each must be managed differently. You can also observe the extent to which wind transports and deposits snow. Snow is widely relocated by the wind and intercepted by vegetation.[1] This deposition is important in that when that snow melts it often provides critically important water to the ecosystems upon which it has collected. But it doesn't have to all be about physical science. Shovelling snow also offers lessons in the humanities.

For example, snow shovelling can tell you a lot about your neighbours. Whether and how often they shovel their snow can be seen as a comment on both their diligence and their sense of place. When I am away and it snows heavily our neighbour will start his snow blower and help my wife with the driveway, which tells you a ton about his generosity and neighbourliness. Continuous snowfall can also tell you a fair amount about neighbours' tolerance. One cold night I found myself out shovelling the continuously falling snow at the same time the same neighbour was shovelling his sidewalk. "I don't mind the snow," I shouted across the street, "but I think I have had enough." "I hate the snow," he shot back, "and I have had more than enough." Okay, I said to myself, don't press that button.

If you persist, however, and don't cave in to the gnawing temptation to go to Mexico or southeast Asia for the winter months as many do, you will observe that the most snow falls

1   Recent research indicates that a great deal of the snow that covers the prairies in western Saskatchewan has been transported there by the strong winds that blow eastward from the mountains and across the plains of southern Alberta. Much has been made of why winds are so fierce in this region. The simplest answer, one southern Alberta rancher told me, was that Saskatchewan sucks.

not when it is coldest, in December and January, but when air temperatures moderate, in late February, March and early April. This is because warmer air can transport more water vapour. As long as air temperatures remain below freezing, precipitation will fall as snow and you won't see any rain comin' down. Light is also critical. I have often argued that I can put up with the cold and snow of the Canadian winter; it is the darkness I find most difficult to endure. After a lifetime in these mountains I have reached the accommodation that if I can make it to February 15, the gradually lengthening days will get me through till spring. By the middle of March incoming solar radiation is surprisingly strong. By the spring equinox the sun has clearly become your friend. It should be noted, however, that intense radiation has a profound effect on water. The water in the human body, for example, is instantly vaporized in a nuclear explosion, and that is one of the reasons why, as we saw with Hiroshima, only a shadow remains where a person might have stood or sat when a such a bomb has gone off nearby.

Over just a few drinks at the end of a water conference in Ottawa, the University of Saskatchewan's Dr. John Pomeroy offered a very memorable scientific explanation of what happens to the water in the human body in a nuclear bomb blast. The radiant heat and radioactivity of the explosion cause the water in the bodies of exposed humans to instantly sublimate from a liquid to a gas. By way of a napkin drawing Pomeroy demonstrated that I should not expect to avoid being vaporized in a nuclear bomb blast by simply hiding beside a snowman. Being Canadian, he quipped, will not save you.

As long as the temperature is not too low or the snow too deep, the spring sun will on its own make the snow go away. It also warms the dark patches on the surface of your driveway or sidewalk after you have shovelled it. This is a good time to begin researching the effects of incoming solar radiation on the snowpack and the ground beneath it. One experiment that always yields greater understanding of snow's response to light is to use snowfalls as a way of determining how deep the snow has to be before its albedo triumphs over the strength of incident radiation. The other way of posing this same question is to determine how warm the sun's rays have to be to defeat the reflectivity of the snow and penetrate deeply enough into it to melt it. Call me lazy, but I have conducted experiments to determine this by shovelling parts of my driveway selectively to see how much exposure was required before the heat building up in the concrete and surrounding rocks would melt the snow off the rest of the driveway without my having to shovel it. I have found that if the snow is deeper than about five centimetres it is usually difficult for the sun to vanquish the snow's reflectivity. Less than that, however, and the cleared strips of the driveway will spread outward like a damn as long as the sun shines. The moment the sun stops shining, though, all bets are off. Melting simply ceases and you may have to get shovelling again before the wet snow freezes. I noticed during the year of this writing – a year in which snowfall was persistent and below-normal temperatures hung on without a break through all of March and early April – that the sun reached a point in its procession into spring at which it penetrated even overcast skies to accelerate sublimation: the vaporization of snow without it becoming water first. This is when even the deepest snowpacks begin to succumb, once again showing the enormous power of direct solar radiation. While not quite requiring as much heat as it takes to turn snow into liquid water, the energy required to sublimate or vaporize one kilogram of snow is roughly equivalent to the amount of heat it would take to raise the temperature of ten kilograms of liquid water 67°C. The fact that sublimation absorbs such a great deal of heat is why it can be cold near snowdrifts in the spring even when the sun is warm.

It is as the spring progresses that the likelihood of rain coming down and falling on snow increases. If this happens late in the season it can mean your snow shovelling duties are over for the next five to six months. If the rain is persistent, however, a chain reaction of sorts begins which can lead very quickly to the scale of flooding that occurred in southern Alberta in June of 2013. The question then becomes this: given that we have lost two full months of annual snow cover in western Canada over the past 50 years, and that winter temperatures are rising – in Alberta, for example, by an average of 5.5°C since the 1960s – should we expect more rain-on-snow events? With further warming should we expect those to occur even in winter as well, as is already happening farther south, in the U.S. Rockies?

Rain-on-snow events should be taken seriously if there are deep mounds of snow still around in your driveway. Care must be taken to ensure that rapidly created meltwater has somewhere to go. Sewer grates must be cleared of snow and ice, and downspouts directed away from the house. If the rainfall persists for more than a few hours, you may want to help your neighbours as well. It will be important to keep updated on weather conditions in case you or any of your friends or neighbours are forced to evacuate.

That is pretty much it for my advice on how to optimize shovelling a driveway, except for a few closing observations. First, always remember that what you are dealing with is water: rain comin' down in a different form. Second, don't forget it can be pleasurable. A clear blue sky following a big snow storm symbolizes what makes winter magical in the northern hemisphere. The sky, the sun and the sparkling snow are a reminder of the glory of being alive on this magnificent Earth. Finally, as the shovelling season draws to an end, know that it is only temporarily over. The snow never really gives up; it merely assumes the form of liquid water or hides invisibly in the atmosphere, where it waits with tireless patience for an opportunity to form and fall again. But know also that the seasons in the northern hemisphere are no longer static. In the absence of snow, the total amount of energy stored in the climate system at any given time will increase. Because it normally covers more than half of the land in the northern hemisphere each year, and possesses such important properties, seasonal snow cover is recognized as a defining ecological factor throughout the circumpolar world. The ongoing loss of snow's refrigerating feedback will cause land surfaces to warm and atmospheric temperatures to rise, with direct consequences for living things everywhere.

THREE
RIVERS OF HEAT

Ed Struzik's 2017 book *Firestorm* is a timely and critical analysis of the forestry, development, public policy and climate change circumstances that now make wildfire a major threat to social, economic, environmental and political stability, not just in North America but throughout the northern hemisphere. In the beginning of the book, Struzik, a nationally respected environmental journalist, introduces the word *megafire*, a relatively new term for fires that burn at least 100,000 acres, or roughly 400 square kilometres. Fires of this size and much larger are now occurring far more frequently, displacing more and more people and reshaping both forest and tundra ecosystems in ways that scientists are only beginning to understand. These are bigger, hotter, faster fires which we are unprepared and ill-equipped to fight and which are causing ever greater damage and human displacement. Struzik makes the case that we can no longer economically afford to ignore the impacts of such large fires on the ultimate sustainability of heavily forested countries like Canada and the United States. Such fires have become as costly as hurricanes and tornadoes.

Struzik worries that as a society we are not taking this rapid and dramatic rise in both the number and intensity of these fires seriously enough. He is concerned, along with many others who have observed these changes in North American fire regimes, that simply putting more money into fire suppression is not enough and will never again be enough. Struzik cites exhaustive scientific evidence that the only fiscally responsible way to address the growing megafire threat is to treat catastrophic fires as the natural disasters they are; to do our best to find better ways of fighting them; and to do more to increase our capacity to restore naturally resilient forests while proactively protecting lives and vulnerable property from future fires by identifying where megafires are likely to occur and acting in advance to reduce risk.

Struzik cites the work of Dr. Mike Flannigan, whose landmark research into changing fire regimes in Canada put the country on notice with respect to the megafire threat and its direct relationship to human-induced climate change. Flannigan has shown that on average about 7,000 wildfires occur in Canada each year. What is interesting, however, is that the area burned has doubled since the 1970s, when global mean temperatures began their relentless rise, especially in northern climes, where the warming is more drastic than in temperate and tropical regions. Flannigan has projected that the area burned is likely to double again by the middle of this century and perhaps triple by 2100.

The situation may actually be even worse farther south in the U.S., where the area burned could increase two- or even threefold by as early as 2050. In addition, in places like Montana, where mean temperatures are expected to rise by as much as 2.7°C by mid-century, there may be an increase of as many as 15 more summer days when temperatures will reach 35°C, or 95°F, which will create drier forest conditions and lead to bigger, hotter, faster fires. Struzik notes that since 2011, more trees have burned in the boreal forests of the U.S., Canada and

Russia than burned in all the rainforests of the world – including the Amazon – during the same period. Adding to the threat is that the conflagrations that are becoming commonplace in California are now advancing northward into the Arctic tundra, where wildfire threatens to accelerate global warming to a possible runaway state if the methane frozen in rapidly thawing peat bogs catches fire.

Flannigan offers three simple things we all need to know if we are to understand why changes in wildfire regimes are going to pose such an increasing threat to our economy and our environmental and political stability in the future.

The first is that the warmer the mean temperature, the drier our forests will become. This is because a warmer atmosphere is capable of holding more moisture in the form of water vapour, which, Flannigan notes, would normally go toward satisfying the thirst of water-stressed trees. More precipitation – if it occurs – can compensate. The key factor is that for every 1°C increase in mean temperature, 15 per cent more rainfall is required to restore the original balance.

The second thing we need to know is that the warmer it is, the more lightning we can expect. Lightning, Struzik notes, accounts for a third of all wildfires in the northern boreal forest and 85 per cent of the area burned.

The third prospect is that the warmer the mean annual temperature gets, the longer the fire season will become.

Struzik notes that in addition to these three factors, there are other linkages that complicate and exacerbate the wildfire threat. These include invasive species, drought, forest diseases and the fact that millions of people live in and enjoy the world's northern forests. Add all these elements together and suddenly the magnitude of the wildfire danger becomes apparent.

The real and perhaps lasting value of Struzik's carefully re-searched and well-written book may be the way he touches on what is becoming an overarching theme of our time, a theme that embraces not just fires but also floods, droughts and hurri-canes. In the absence of integrated plans for dealing with such emergencies, and the lack of acceptance of the link between such disasters and accelerating human-caused hydro-climatic change, we are left with relying largely on heroic emergency measures to protect and rescue us. In effect, we are leaving it to the heroics of individuals to save us from our collective wilful blindness and folly.

At present, we as a society are still refusing to even admit the inevitability of mega-wildfires, never mind accept the things we must do if we want to be resilient to this and other threats related to climate change. Year-round and runaway fire seasons are now beginning to happen. We have to learn to live with that. I should point out that all of this is covered in just the introduc-tion to Struzik's book.

From this general analysis of the megafire threat, Struzik goes on to name names and reveal failures in the battle to under-stand and react to the ways our population's needs and num-bers, together with the climate change effects we are creating, are altering the fundamental conditions that make megafires not only possible but inevitable. Struzik points to policies in both Canada and the United States that have resulted in fund-ing being diverted from prescribed burns and other proactive forest management practices to fire suppression. This is far more important than it might appear. It means that as a so-ciety we have decided to fight fires rather than prevent them, which in an expanded sense is what we are also doing with the broader threat of climate disruption.

Struzik cites the western Canadian province of Alberta as an example. Despite clear and growing evidence of increasing vulnerability to megafires, collected between 2002 and 2016, the government under Premier Rachel Notley cut the province's wildfire prevention and management budget. Why? Because of rising provincial debt accrued largely by previous governments. At the same time, Struzik reports, British Columbia's premier, Christy Clark, and her Saskatchewan counterpart Brad Wall supported a renewed effort to create a national fire strategy, but at the same time refused to accept that anything further needed to be done about climate change. As Struzik mentions, Notley was sincere in her concerns about climate, but not at the expense of the political fallout that would have resulted from standing in the way of oil sands and other fossil fuel production so vital to the province's economy, the very economy that is driving the changes in climate behaviour that make megafires inevitable now and will even increase their frequency in the future. The great irony in this, of course, is that the national wildfire strategy being advocated in Canada did not come into existence because of these factors. In other words, petroleum interests derailed a strategy that would have saved lives and property.

As Struzik points out, the first victim of these policy failures was the northern Alberta city of Fort McMurray, the very centre of oil sands production in the province and a controversial contributor to the climate impacts that later fed the megafire that very nearly destroyed the town. What we learn from this is that fire is indifferent to money. One of the points Struzik reminds us of is that the fire made three runs into Fort Mac before the order to evacuate was made. Here it is – and it tells us so much: the evacuation was delayed because no one wanted to slow oil sands production. The symbolism inherent in these decisions is impossible to ignore. The metaphors snap to attention and pile atop one another like sedimentary layers over an oil deposit.

As Struzik notes, the fire risk at the time was orders of magnitude above normal, yet no limits were imposed on all-terrain-vehicle use in the tinder-dry forests of the Horse River where, it is surmised, the fire was started, whether by someone carelessly tossing a cigarette butt or by dry grass or twigs coming into contact with the hot exhaust pipe of an ATV. It was as if no one took the fire hazard seriously and no one wanted to acknowledge the degree of the threat, even when parts of the town started burning, for fear of the fire shutting down local commerce or impeding the magic multi-billion-dollar machine that miraculously turns tar into new vehicles, houses, and holidays in the sun. Astonishingly, no one appears to have known that the fire had entered the city until citizens started tweeting one another and sending photographs from their smart phones. It appears from the evidence Struzik puts forward that no one had anticipated how hot the fire was and how fast it was moving. Homes were suddenly engulfed in flames 15 metres high. A motel burned completely to the ground in only 45 minutes. Fences caught fire and barbecue tanks started exploding. Firefighters didn't know whether to put out fires that were burning down houses or try to prevent the fire from engulfing what wasn't already ablaze.

Because provincial firefighting efforts were coordinated on a different radio frequency, local firefighters were unable to request and direct air tanker support. With so much on fire at once, all that local firefighters could do in many instances was assist in evacuation of residents, which in itself was a chaotic effort that succeeded, surprisingly, not because it was well orchestrated but because of the sheer commitment and utter heroism of emergency services personnel, in particular the RCMP, firefighters and municipal staff. Even at that, however, Struzik reports there were jealousies over who was in control, issues of conflicting jurisdiction and serious failures in communication, just as there often are in almost all natural disasters. We can expect more of these same failures as the climate change threat continues to accelerate.

One question that was not brought up, at least publicly, during the Fort McMurray fire was what would have happened if one of the open-pit oil sands operations had caught fire. Struzik reports that some oil sands companies took independent action to protect assets when two of the largest blazes around Fort Mac merged to create a firestorm over more than 250,000 hectares. Firefighters from South Africa, Mexico, the United States and nearly every Canadian province and territory were brought in to help extinguish the blaze. It must therefore have occurred to someone that if any of the oil sands operations had caught fire it would have been difficult if not impossible to extinguish, perhaps for years, if ever. My justification for making this claim comes from a conversation I had, during my last visit to Fort McMurray, with a technician operating one of the extraordinarily sophisticated command centres that monitor every step in the mining and processing chain at one of the big oil sands facilities. When asked what the biggest threat to the safety of this massive operation was, he did not hesitate even for a moment in saying "fire," which makes sense given they were mining millions of tonnes of combustibles. That question, in a changing fire regime, needs to be publicly answered. What *would* happen if an oil sands mine were to catch fire? Could such a fire even be extinguished? Does anyone actually know?

Struzik explains that the hot air rising from big fires creates what are called pyrocumulonimbus clouds. These fire clouds are capable of generating lightning that can start fires in areas surrounding the main fire. What is interesting about the behaviour of these clouds is that the heat of an intense fire creates an updraft which sucks smoke, ash, burning twigs, branches and flaming treetops into the sky. The heat of the fire also sucks water vapour from lakes and streams into the smoke and everything burning in it. This smoking mix of fire and water then cools and assumes the shape and behaviour of a thundercloud, with the flaming materials and hot water vapour adding significantly to the energy in the cloud. In rare cases, rain produced in these clouds can help put out the fire below, but in most instances the heat and particulates in the smoke trigger a chemical reaction that prevents precipitation from forming.[2] What is formed instead is lightning, which can play havoc with the fire beneath the cloud and cause more fires in surrounding areas, often a great distance from the main fire. Struzik reports that in extreme pyrocumulonimbus situations, as many as 8,500 lightning strikes have been recorded in just one day. In really hot fires what can happen is what Struzik describes as meteorological chaos. The cooler air rushing in to replace the rising hot air creates high winds, often with powerful gusts

2   This often happens also in smoky conditions downstream of large fires. Raindrops are formed around dust motes called nucleates. In smoky conditions there may be so many nucleates in the air that even with moderate amounts of water vapour present, the drops that form around each suspended mote of ash do not become large enough to fall as rain. That is why rain predicted in these conditions does not materialize or the rain that does fall evaporates before it reaches the ground.

that can cause the fire to move in unpredictable directions, which is exactly what happened with the Fort McMurray fire. In these circumstances, Struzik explains, updrafts can suddenly collapse, adding to the unpredictability and spread of the fire. When updrafts collapse, downdrafts can bring burning treetops, branches and embers back to the ground, spreading the fire in all directions. This can enable a big fire to advance as much as two kilometres in a matter of hours. Some fires, Struzik notes, have travelled as far as 60 kilometres in just ten hours.

Until recently it was held that pyrocumulonimbus clouds, like the flat-topped thunderheads they resemble, did not possess the energy to break out of the lowest layer of the atmosphere – called the troposphere – to enter the less dense but much more mobile stratosphere above it. It was held that only the very largest of volcanic eruptions were powerful enough to pierce the stratosphere. Struzik, however, cites recent findings that megafires can generate enough energy to enable pyrocumulonimbus clouds to pierce the stratosphere, and for the ash and debris carried that far aloft to circulate globally in the upper atmosphere in the same way ash from big volcanoes does, with the same effect, albeit with lesser potential short-term influence on the global climate. As Struzik points out, it is not hard to see why megafires constitute barely 3.5 per cent of all the world's wildfires but are responsible for 95 per cent of the fire damage.

The way things stand now, with governments still concentrating disproportionately on fire suppression and emergency heroics as opposed to preventive management, we should not expect the megafires situation to change in the immediate future. In fact, we should expect things to worsen and even become permanent. But what is not being understood here is that as megafires become more common and more of them attain the stratosphere, smoky skies could become a global phenomenon. In other words there will be fewer places on Earth where the sky will be as blue as we remember it.

Struzik again cites Dr. Mike Flannigan, who has predicted that not only is it likely to be warmer and drier in the boreal in future, but there will be a lot more mature trees in the forest. This proliferation is due to the extent of fire suppression since 1950, when governments decided that instead of mimicking natural processes of fire suppression through more frequent lower-temperature burns, all wildfires should be extinguished as soon as possible. With half a century of Smokey the Bear hitting the public over the head with a shovel every time forest fires were mentioned, it has been difficult to accept and facilitate the role of wildfire in forest succession, especially in a changing climate. Accommodating the natural role of wildfire in healthy forest succession is no longer an option, however, if only for the simple reason that we can no longer afford the status quo. Struzik reminds us that before the megafire of 2016, Fort McMurray was considered to be at only medium risk from wildfire. There are now many communities in Canada and the western United States that are at much higher risk than Fort McMurray was assumed to be in 2016.

Struzik reports that in Canada, fires larger than 200,000 hectares occurred only four times between 1970 and 1990. Since then there have been 12 of that magnitude. In the U.S., fires of 200,000 hectares occurred only three times between 1983 and 1999. But between 2000 and 2016, there were 11. Struzik also reports that the cost of fighting wildfire in the U.S. rose from $240-million in 1985 to $2.1-billion in 2015. In Canada, annual firefighting costs now top as much as $1-billion and are expected to continue rising. Where I live, in Canmore, Alberta, wildfire is not an abstract threat. Our house is adjacent to forest that has not burned in decades. Our backyard, which for ecological purposes we have left wild since we built the house 25 years ago, is dense with lodgepole pines that in some cases have to be more than a century old. I am reluctant to cut them down, for aesthetic reasons, but I am fully aware of the growing risk from fires. I can say this because ten years ago I actually set one.

All those years of listening to Smokey the Bear had deeply affected me. "Only you can prevent forest fires" was a mantra I have never been able to get out of my mind. But I had a good reason to set this forest on fire. As I explained in my 2010 book *Ecology & Wonder*, it was the United Nations International Year of Mountains. During that year we were trying to learn as much as we could about mountain ecosystems. I had taken a course from fire and vegetation specialists in the mountain national parks which taught that fires were important to natural plant community succession in mountain places. We were also taught about fireproof clothes and drip torches, soil moisture and wind direction. More importantly, we learned about firebreaks. It is important to note that I was also under the close supervision of experts who indicated to me that I might feel differently about fire as result of starting one. They said they themselves had felt something almost primal about setting forests alight, as if in the memory of our species we had done this before, and often. They even predicted I might find the experience deeply satisfying. "It might even awaken your inner pyromaniac," joked one of them. They were right.

Not everyone has the occasion to purposely start a forest fire. This one, however, had been planned for months. The area where the fire was scheduled to be set is a unique part of

Banff National Park. Known as the Fairholme Environmentally Sensitive Area, it encompasses the rich montane lowlands of the Bow River Valley at the eastern edge of the park. It is some of the best winter habitat on the eastern slopes for elk and deer and one of the last remaining places in the region that can sustain relatively stable predator populations. The purpose of this fire was the further improvement of wildlife habitat so that the area would continue to support a viable wolf pack, a sustainable local cougar population and a few more black bears.

Ian Pengelly was in charge of the fire. As a fire and vegetation specialist for Parks Canada's Banff National Park field unit, Pengelly had a lot of fire experience. He had waited a long time for conditions for the fire to be perfect. As snow had fallen the previous week, the soil was perfectly saturated. The temperature was right and the winds were light.

Upon instruction from Ian, provincial fire specialist Terry Studd and I descended to the bottom of a small ridge and began to set fire to clusters of south-facing junipers. The fire quickly rose up the slope, just as Pengelly had predicted it would. I took over the drip torch from Terry as we advanced along the base of the ridge. The gear we were using was the Western Forester Seal-Tite Back Fire Torch. When sealed, it looks like a big chrome coffee carafe. A large metal cover screws into the top. When this is removed one finds a metal wand that can be screwed back onto the outside of the tank to create a device that can literally change the world.

By opening the right valves and lighting the end of the wand, one can create liquid fire by simply pouring the fuel in the same way that you pour water from a watering can. I watched Terry as he advanced along the base of the ridge and made his way up the steep slope to meet a fire line created by Pengelly and Tom

Davidson. He looked like a pied piper from hell. As he moved nonchalantly up the hill and through the forest, flames popped out of the ground and followed him. A great roaring followed the fire into the forest above.

As the smoke cleared, the afternoon winds stopped blowing from the west. Cooler, denser air began pouring slowly down the mountainsides into the valley. Night would soon put the fire to sleep. Pengelly explained that Parks Canada had to be very careful to keep its fire program operative within constraints acceptable to the local residents, upon whom the agency relied for support. Ian was very conscious that the smoke from prescribed burns in the park could create discomfort among people with respiratory problems who lived downwind. For this reason Pengelly and his colleagues carefully monitored conditions and restricted each burn to less than 200 hectares. A firebreak had also been created between the park and the neighbouring communities of Harvie Heights and Canmore. Pengelly hoped the people who lived downwind in the Bow Valley would appreciate that the Fairholme fires would ultimately contribute to the natural biodiversity of the park and the region, making it a safer, more interesting and ultimately more worthwhile place to live. As I prepared to leave at the end of the day, I thanked Ian for one of the most important and instructive experiences I had ever had in the mountains. I promised, however, that I would keep my newly awakened inner "pyro" in check. He waved goodbye, then went back to check the fire "one more time."

This would not be my only opportunity to learn from experienced Parks Canada ecologists like Ian Pengelly. In order to fully appreciate Parks's commitment to restoring the ecosystem function of fire in the Canadian Rocky Mountain Parks World

Heritage Site, I was later invited to fly over the part of Banff National Park where fire specialists had been most active in re-producing natural fire regimes through prescribed burns.

After we lifted off, Pengelly made a very telling observation on the relationship between human populations and ecosystem dynamics. Horses and smallpox set the stage for mobility and mortality that altered human influence on the landscape. It was this that led to the fuel buildup that fed the big railway-era fires. In the absence of people, the forest aged and dead branches and fallen trees created the perfect conditions for huge blazes. This rang a bell. Recent research suggests that the planet's atmos-pheric carbon dioxide levels fell after the great plagues of the Middle Ages, as forests expanded in the absence of humans and a great number of trees absorbed and captured more CO_2. There is a strong and direct link between people and forests. It is almost as if the one presupposed the existence of the other.

We left the Banff warden office and flew down the Cascade River Valley to the Red Deer River and then west to the Pipe-stone Valley, north of Lake Louise. Pengelly described the fire history of this region of the Rockies and tied it to management decisions made in the past. As he pointed out the small patches of forest that had been set ablaze in prescribed burns, it oc-curred to me that despite the huge cost, both in science and in the efforts that will continue to be necessary if we want to even begin to duplicate natural fire regimes, there is still a great deal we need to know and do to fully understand and reproduce the effect of wildlife on the landscape of the mountain West. Im-portant work is being done here that will help everyone in the region better understand and control fire and its effects. This knowledge will help us slow the diminishment and loss of eco-system vitality, and help us delay some of the climate change effects that have already begun. But it won't keep the pine bark beetle out and it won't stop the flow of invasive species into the World Heritage Site. Even fire ecologists like Pengelly would agree that we are too timid and still too inexperienced to do what nature once did for us for free in the forests of the mountain West.

Flying from Lake Louise down the Bow Valley back to Banff, Pengelly didn't need to say a word. It was perfectly obvious, looking at the unexpected extent of forest that hasn't burned in more than a century, that we haven't done enough to restore fire's former role in this ecosystem. We haven't even begun to burn what we need to burn. I think of Banff's conservative, profit-obsessed business community and I wonder which would scare them more: the size of the forest in the upper Bow Valley that should be set ablaze to maintain natural plant succession, or the extent of the vulnerability of the Banff townsite to a catastrophic wildfire fuelled by deadfall that has accumulated in a forest that hasn't been allowed to burn in a hundred years. From the air it was easy to see how little actually protects the town of Banff from an up-valley fire or one that might leap into the valley from any one of a half dozen side valleys and adjacent passes. Skilled firefighters might, if there were enough warning, be able to turn a big fire to the rims of the wide valley at Moose Meadows near Castle Junction. But then again, with a strong wind from the west, they might not. What was needed was a firebreak in the vicinity of Castle Junction.

After the helicopter touched down at the warden office in Banff, Pengelly made a final observation on the unexpected risks associated with not allowing fire to play its natural role in shaping ecosystem dynamics. It had to do with the wind. It was his experience, gained from a couple of decades of starting and controlling prescribed burns, that there was usually very little wind directly down the main valley from the direction of Lake Louise. When it was windy in the Banff townsite, he said, the winds usually came down from adjacent Healy Creek. A big fire from that direction, he mused, might be turned at Vermilion Lakes so that it skirted the town on two sides. But it would be tricky. The valley below Healy Creek is a natural venturi that would concentrate wind and thus accelerate it. A hot fire pushed down-valley by the right winds could easily burn the town of Banff to the ground. There was just enough wind blowing from the west that day for me to clearly understand Ian's point. A higher-energy atmosphere will be one of the consequences of climate warming. We should expect higher winds and more frequent and intense weather events. With these changing conditions, the chances of a "perfect storm" in the Bow Valley will be much increased.

While fire ecologists like Cliff White and Ian Pengelly did an excellent job of pioneering prescribed burns in Banff National Park, and in so doing created a firebreak between the town of Canmore and the high-risk areas Struzik identifies in his book, I do not doubt for a moment that both Banff and much of the community where I live could burn down if a big fire came roaring down the Bow River Valley, or if something got started up the Spray River Valley that, fanned by high winds, could not be quickly extinguished. As Cliff White has often stated publicly, what we have created here in terms of forest dynamics is not natural. We have concentrated tens of thousands of people in a place where their expensive homes have a high probability of burning, a fact we downplay for fear of alarming people or threatening property values. White made these observations fully ten years before this writing.

Fire ecologist Marie-Pierre Rogeau explains that fire has always been a threat in mountain and boreal forests. She has carefully documented the extremely large wildfires which burned frequently during decade-long droughts in the recent past. Tree ring data from the 1700s, for example, reveals 14 significant drought years. Rogeau surmises that by the end of the century, much of the mountain West would have looked like a moonscape. Only persistent fire refugia protected by topographic features could have resisted these fires. Millions of hectares of forests in the mountains and foothills of southern Alberta were also lost to burning between 1861 and 1870, as well as in 1889, 1910 and 1936. These important fire years, she notes, also extended into Montana and other western states. Timber surveys undertaken in southern Alberta shortly after the big fires of 1910 revealed that only about 20 per cent of the forest was evaluated as being older than 60 years. Once again, the forest mosaic at that time was a legacy of five significant drought years in the 1800s and 1910.

Dr. Rogeau points out, however, that fire is a disturbance natural to many forest communities around the world, and that a number of studies have investigated how recent burns can in fact create resistance to future fires. When there isn't much to burn, the intensity of burning is reduced and fires die out quickly in young forests. On another positive note, insect populations also get eradicated in fires. If refugia are numerous, they stimulate the forest to regenerate itself.

Rogeau further notes that during a 2017 fire in Waterton Lakes National Park, a nearby weather station recorded a temperature of about 50°C as the fire roared past. These are the kinds of temperatures that are required to open what are called serotinous cones. Many species of pine, including our local lodgepole, have evolved this type of cone. Serotinous cones are covered with a resin that must be melted for the cone to open and release seeds. When a fire moves through the forest, the cones open and the seeds are distributed by winds and gravity. So evolved are our mountain forests that they begin seeding themselves even as a fire burns through them.

Rogeau reminds us that fire and climate specialists have been predicting and anticipating for decades that it would be unlikely that global warming will be kept to less than 2°C. She says the positive feedback loop generated by an increase in temperature has many potential ramifications, including unforeseen effects on the explosion of insect populations, as exemplified by the mountain pine beetle, which has defoliated vast areas of forest in western North America. Rogeau also cites the effect of fire suppression on the accumulation of fuels and the lengthening of intervals between fires, which in turn, in conjunction with serious droughts, are leading to extremely severe burnoffs where no organic matter is left and there are no seeds nor even any suitable seedbed for vegetation to recover. And to make matters worse, forests will struggle to regenerate and become efficient carbon sinks. Rogeau is also concerned about permafrost thaw, due to the severity of the burns and to fires that become self-fuelling as a result of the methane they release. Rogeau agrees with Cliff White and Ian Pengelly that carbon emissions are far greater when mature forests burn than occurred in more frequent, lower-intensity fires that were common in many forests prior to the decision to fight all wildfire. Like White and Pengelly, Rogeau favours small-scale, carefully planned and executed prescribed burns as a means of reducing the risk of megafires.

Ed Struzik argues that the only thing we don't know about

the growing threat of megafire in Canada is where the next one will happen. There is no question in my mind that one of the places it could happen is right where I live. When we are talking about precious forest refugia in a time of rapid hydro-climatic change, this is no small matter. If the upper Bow Valley is a true refugium, because of its UNESCO World Heritage Site designation, then fire must be a serious ongoing issue, critical to sustaining the century-old international tourism economy of the region certainly, but also to protecting the billions of dollars of privately and publicly owned infrastructure that has been built here. In other words there are plenty of reasons to pay attention to changing fire regimes.

One of the realizations that is largely missing from popular understanding of the bigger, hotter, faster and more frequent fires Struzik describes is their negative legacy. The impacts of megafires live on for years in terms of diminished water quality in rivers flowing from and through burned areas; in the lingering effects of compromised air quality on human health; the harm to wildlife; and damage to the psychological well-being of people whose lives have been devastated by disaster. Finally, there is the most dangerous threat of all – and the one we understand the least: the potential of megafires to ignite so-called "carbon bombs." Carbon bombs are warming-induced carbon releases from what were previously major carbon sinks held in place by permafrost and long, cold winters. When methane, for example, is suddenly released from thawing permafrost it can cause wildfires to burn in ever more dangerous ways.

What is of growing interest to me are the hydrological effects of fires. The soils in fire-baked forests change. Chemical compounds that get vaporized, such as the waxy coatings on the leaves of certain plants, are driven into the soil by the heat. As the fire passes, these compounds condense, forming a layer just beneath the surface that is largely impervious to water. These so-called hydrophobic soils are unable to absorb subsequent rainfall. When even moderate rain occurs, millions of tonnes of ash, woody debris, heavy metals freed from the wood by the heat of the fire, and nutrient chemicals such as biologically available phosphorus are flushed in what are often floodwaters through affected areas and downstream into unburned parts of the watershed. This often has highly damaging effects on water quality and on the health of aquatic ecosystems, which are often already compromised or diminished for other reasons. The hundreds of tonnes of sediments mobilized in this way can fill lakes and reservoirs and make water treatment in towns and cities more complicated and expensive for several years after a big fire. The effects of a hot fire can also cause groundwater movement to slow or cease. The loss of the natural capacity of soils to absorb rainwater and snowmelt adds to the likelihood of flash flooding. These impacts can be not only long-lasting but permanent.

Research has also revealed that phosphorus levels can, depending on soil type and other factors, remain very high for a very long time after a particularly hot fire. As a result of these compound and cumulative effects, water temperature in streams and rivers can rise, with serious effects on aquatic ecosystem recovery.

From all this we see that, depending on the state of a given watershed prior to a megafire, big, hot burns can affect forest ecology, wildlife populations, soil recovery, local and regional hydrology, water quality, air quality and public health. Megafires of sufficient extent and temperature can also result in watershed degradation. All this even before a wildfire burns down a town or city.

Struzik also notes that there is a completely unanticipated link between megafires and the lingering impacts of abandoned and orphaned mines. It has been discovered, for example, that in forests near abandoned asbestos mines, asbestos will likely have accumulated in trees growing in the polluted air and soil. In bigger, hotter fires the asbestos contaminants in the wood suddenly shoot out of burning trunks and branches, proving once again that ignoring where pollution goes will inevitably come back and slap you in the face. As Struzik says, the health hazards associated with asbestos have been well known since Roman times. This pollution doesn't go away. In the case of lingering fallout from mines that may have closed decades ago, asbestos needles can even persist in the dead bark of fallen or felled trees where vermiculate fibres have accumulated. It is a health hazard to burn firewood that is contaminated in this way, because combustion can introduce these almost indestructible asbestos fibres into homes wherever such firewood is burned. The stuff can even linger in the ash that remains in fireplaces, potentially reversing many of the gains in the protection of human health brought about by the banning of asbestos as home insulation.

In big fires that happen to occur near asbestos mines, the same asbestos needles that emerge from trees in the heat of the fire are also transported in the form of ash and fiery debris wherever the smoke blows. This, Struzik notes, presents a huge risk to firefighters and to everyone who lives where the ash falls. The same problem exists around abandoned and orphaned gold mines. I have witnessed this personally at the Giant Mine site near Yellowknife in the Northwest Territories, where the risk is not asbestos but a far more immediately dangerous poison, arsenic.

Mining gold requires the careful management of hazardous by-products. Arsenic exists naturally in the hard rock of the Canadian Shield. As a result of milling to separate the gold from the rock in which it is embedded, arsenic trioxide becomes concentrated in the tailings. Arsenic trioxide is one of the more poisonous substances on Earth, and it also happens to be water soluble, which can present serious problems with groundwater. Milling also involves "roasting" to melt the gold out of the ore. In the early years of the Giant Mine, the arsenic trioxide simply went up the smokestack and was rinsed off the surrounding rock by rain and snowmelt. Soon, however, every living thing that relied upon anything growing around the mine site died. Scrubbers were put on the roasters to take out the arsenic trioxide, which was bagged and stored in old tailings and mined-out stopes. But soon there was so much of the deadly poison that special below-ground storage chambers had to be built. The chambers were protected by 5 to 100 metres of surrounding permafrost. Some 270,000 tonnes of arsenic trioxide were disposed of in this way, enough arsenic, as one local water expert joked, "to kill every man, woman and child on Earth three times over."

The engineering thinking behind this solution is interesting. It was held that the permafrost would reclaim the stopes and storage chambers, encasing and freezing the arsenic dust in ice. This was thought to be a brilliant solution. Not taken into account, however, was the amount of heat entering the chambers and stopes from the mine workings, eroding the permafrost from below. Another consideration omitted was that the arsenic trioxide dust was still hot from roasting when it was bagged and stored, which caused permafrost melt to radiate outward from the stopes and chambers. The melting was further exacerbated by a decision to permit open-pit instead of underground mining of the ores. The open-pit system began to erode the remaining permafrost from above. Milling of ore ceased in 1999, and in 2004 the company that owned the Giant Mine declared bankruptcy and walked away from the problem. But the permafrost did not stop melting. If fact, with climate change it is accelerating.

What exists now at the Giant Mine is a series of storage chambers and old stopes containing enough arsenic to poison the entire Mackenzie River system from Great Slave Lake to Inuvik. The only thing that presently keeps that poison from becoming mobile in groundwater is the fact that the dust and its surroundings are only a few degrees below the freezing point of water. The people of the Northwest Territories have been left holding the bag. The governments of NWT and of Canada are now faced with having to artificially refrigerate the stored water-soluble arsenic trioxide to prevent it from being mobilized by groundwater. While the land surface around the abandoned mine and even the tailings can be remediated, the stopes and storage chambers and their surrounding permafrost will require constant artificial refrigeration and monitoring at huge public cost, probably forever.

The threat of megafires, Ed Struzik explains, has revealed yet another threat. Preliminary research has shown that toxic levels of arsenic have accumulated in the forests around the abandoned mine. Little is known, however, about the extent of arsenic uptake that may have occurred in plants and particularly in trees. Neither is it known whether arsenic residues are redistributed by wildfire. One concern is that if there is uptake of arsenic into forests around orphaned and abandoned mines, hotter fires could result in the release of highly toxic arsenic gas, which would be yet another serious hazard facing those charged with firefighting. We don't know, and probably won't know until there is such a fire. What we do know, though, is that there is a risk. Struzik reports that perhaps as many as 10 per cent of the 100,000 to 500,000 small to medium-sized abandoned hard-rock mines in the United States have been identified as threats to public health and safety. There are also some 10,000 abandoned mines in Canada, many of them located in the northern boreal forest, the public health and safety status of which has yet to be assessed. To put it simply, there is much we don't yet know about what the growing number of megafires will mean in terms of the redistribution of chemical contaminants lingering in the ground or in the wood of trees, or that have been transported through the air or into our water regionally and globally.

One thing the new focus on the effects of megafires has done is provide a whole new perspective on smoke as a human health hazard. In the same way as water is an almost universal solvent, our planet's fluid atmosphere appears to have an almost unlimited capacity to absorb and transport whatever substances high temperatures or combustion can vaporize. The number and range of items in this category is astonishing. Struzik reminds his readers of the number of harmful substances that are released when it is just tobacco being burned. When you light a cigarette and inhale the smoke, you are not only sucking tar and nicotine into the wet interface between the atmosphere and your living body, but also breathing in benzene, a chemical known to cause a variety of cancers. But that is not the only carcinogen you are inhaling. You are also breathing in cadmium, an element responsible for damage to the brain, kidneys and liver; lead, a heavy metal known since the beginning of civilization to cause brain damage; and mercury, another heavy metal that bioaccumulates in the food chain, with disastrous effects on human health. Struzik stunningly reports that as many as 600 substances begin burning when you light up a cigarette or a cigar. But that is not the end of it. The burning of those hundreds of substances results in new chemical reactions which in turn create up to 7,000 other compounds, many of which are poisonous. Beyond the known link between smoking and human illness, we have little idea how long these poisons persist in the environment, or whether they are absorbed and released again into our air or waters. You can see where Struzik is going. If this is what happens when you light a cigarette, what happens when you burn a forest?

Imagine the smoke from the simultaneous burning of thousands if not millions of trees in a megafire, to say nothing of the hundreds of other plant species that live on the forest floor, and the chemical cocktail that is created in such a firestorm. It is hardly any wonder that, in a sky filled with smoke and poison, the sun becomes a red rubber ball. Now imagine living in parts of China where the sun rises as a red ball in the morning because of so many pollutants in the atmosphere. Then imagine the extent of smoke in the sky never diminishing. The contaminants to which most Chinese and many others in Asia are subjected as a matter of course in their daily lives include toxic ash, ozone, carbon monoxide, free radicals like formaldehyde and acetaldehyde, and mercury. It is like inhaling fire smoke, only worse. Living in parts of China is like living perpetually in the smoke of a megafire, a great incendiary not just of trees but of everything we have to burn in the world to sustain our current prosperity.

Because of the huge population of China and the size of its economy, it is not unreasonable to say that the Chinese have set fire to the sky, and that this fire could make parts of their country uninhabitable at least for parts of the year as temperatures rise globally. It also threatens to bring down Earth System function regionally, which would have huge potential effects on the stability of the larger, biodiversity-based, self-regulating planetary life support system upon which our economic and political systems also depend. In other words, the fact that China has set fire to its own skies, turning the sun red for weeks on end, is a threat to us all. It is like subjecting all of us, in any place where pollution originating in China ends up, to the effects of a lifetime of smoking even if you have never lit up a single cigarette. And these effects are not limited to air. They also affect water precipitated out of the air as rain or snowfall, which through the intermediaries of soil and forests finds its way ultimately into groundwater and then surface water, not just in China but around the entire planet. Now imagine the additional threat of megafires across the northern hemisphere adding to the pollution China is creating. Now imagine turning up the heat in the oven that is our global atmosphere and you can see where we are going. What we are facing now is nothing less than a global atmospheric and hydrologic emergency.

As a consequence of the warming caused by the myriad ways humanity has altered the composition of the Earth's atmosphere, we are learning a great deal more about the long-term effects of deep and persistent drought and the links between drought and wildfire. One thing we have discovered is that trees cannot benefit from higher temperatures or a greater concentration of carbon dioxide in the atmosphere if they don't have enough water to take advantage of these conditions.

In *Firestorm*, Ed Struzik cites Canadian Forest Service research that shows that the browning of vast areas of white spruce forest in the Fort McMurray area may have been caused by drought that has plagued the Canadian boreal since the beginning of this century. This extended drought has weakened the trees, making them more susceptible to insect pests and disease, which in turn makes them more vulnerable to wildfire. While trees will, over time, move upslope or northward to escape persistent drought, they often run out of room. As Struzik points out, no forest type is invulnerable to future droughts. What is even more important to understand is that under changing climatic conditions that favour accelerated and expanded fire regimes, the result in time will be the emergence of ecosystems that don't resemble those the fires displace. We can now expect fires so hot that little grows afterward, because burned soils are so damaged that seedlings can't take hold and root. The fact is that, at present, we are nearly helpless to do much about the megafire threat we have created for ourselves.

From Struzik's important book there are six lessons that stand out. The first is that not all natural wildfires are the ecological disasters they may at first appear to be. That said, increased wildfire frequency and intensity will alter patterns of forest recovery in a climate-changed world. The greatest impacts will be on watersheds, public health and safety and the economy. Business as usual is not an option. We need to halt watershed degradation and protect and restore full watershed function. Most importantly we need re-establish the long-standing tradition begun and maintained by Indigenous peoples of setting frequent, regular, controllable fires before the snow melts in the early spring and as temperatures drop in the autumn. Such fires enhance wildlife habitat while at the same time reducing fuel buildup, and in so doing they dramatically reduce the threat of bigger, hotter, faster fires that destroy soils for decades and threaten people and property. Finally, we need to see wildfire – water's diametric and symbolic opposite – as an integral part of a healthy forest, and nest that realization within the larger promise of a restoration imperative.

The good news in this story is that over the ten years since I was invited to participate in that prescribed burn in Banff National Park, Cliff White and his colleagues Ian Pengelly and Mark Heathcott and others have continued the campaign for frequent, early-season prescribed burns to reduce fuel load in the valley where I live. Though not perfect, expansive fire-breaks now exist that reduce the risk of a megafire that would threaten the towns of Banff and Canmore. These programs need to be maintained and expanded, however, and as White explains, individual communities have to take fuller responsibility for wildfire protection by supporting regularly scheduled prescribed burn programs outside and within their boundaries, and by creating a "firesmart" culture among citizens and visitors. But even Cliff White cautions that in a warming climate wildfire is a growing threat, a threat we will have to learn to live with.

FOUR

RIVERS OF WORDS

My dear Merrell-Ann,

I was deeply troubled by the book you recommended, as I was meant to be. Author André Alexis did not mean *Fifteen Dogs* to be taken lightly. I have to admit, in fact, that I felt you purposely passed on to me an intellectual bomb that you knew would go off in my upturned face. But as is the case with the best of such time bombs, it went off at just the right time in my intellectual life.

I find it interesting that the protagonists in the book are either gods or dogs, which in itself is a stunning play on words in that as the novel progresses it is sometimes difficult to tell if one isn't the inverse of the other. Of the gods, Hermes is of greatest interest to me. Hermes, god of herds and flocks, travellers and trade, thievery and cunning, heralds and diplomacy, and language and writing, as well as astronomy and astrology. It could be said that his extensive portfolio is remarkably similar to mine at the UN. It sometimes seems to me that all I do is travel, deal with herds, fight off thieves, arm-wrestle diplomats, and finally write and prognosticate on the future by pondering the alignment of the scientific and political stars. Hermes's duties extend far beyond mine, though. Hermes is also a personal messenger of Zeus, as well as being guide of the dead, the god responsible for leading departed souls into the underworld. That is a job for which no mortal can apply.

Despite these onerous responsibilities, however, it is Hermes – and not Apollo, the god of the sun, the light, music and prophecy – who demonstrates the keenest appreciation of irony and the best sense of humour when it comes to observing the nature of the world of mortals. Hermes also knows and cares about water. You will remember that at one point he and Apollo meet to discuss the bet around which the plot of the novel turns. The setting of the meeting is the Wheat Sheaf pub in Toronto.

> The bartender, a devout young woman, approached, her head bowed, unable to look at the gods directly.
> "Is there anything I can get for you?" she asked. "Anything at all? I would be honoured."
> "I like this Labatt's," said Apollo. "Give me another."
> "You like it?" said Hermes. "It's a waste of perfectly good water."
> "Philistine!" said Apollo.

Hermes – my kind of guy.

The plot of *Fifteen Dogs* is riveting. On one side, the immortals. On the other side, us. In the middle, the dogs. The immortals can no more understand what it is to live with death than they could grasp what it was to live without it. This is what is fascinating to the gods about us. Death is, as Alexis points out, in every fibre of our being. It is hidden in our languages and is part of the foundation of our civilizations. It can be heard in all the sounds we make and in the way each of us moves. It darkens our pleasure, but it also lightens our despair. We know that, whatever we are going through, in the end we will die.

Hermes, the god of translators, the god of interpreters like me, understands that keeping our language is far more important than we may know. It was the American writer William Kittredge who made me aware of the fact that we have our own language in many parts of the mountain West. To preserve what is essential about where and how we live, we must preserve this language. Kittredge and Alexis's Hermes would be good together at a party. Both acknowledge that the loss of words can lead to the loss of the things those words stand for. As Kittredge points out, the devaluation of words makes for the devaluation of the things words describe. A vicious circle is created from which there is no escape. With fewer words to describe the places that surround us, it becomes harder to justify saving them. As these places vanish from our direct experience, the need for a language to describe them vanishes as well. Kittredge warns us that languages can decay and die. Once the language that people use to tell the story of who they are vanishes, the sense of self can be lost. People can become less than they were. The same thing, Alexis might argue, happens to dogs.

Kittredge maintains that we must be careful to preserve local names and stories. In his *The Nature of Generosity* he says that

> naming helps people witness themselves and reflect on what they've seen. It is the beginning of talking to ourselves, that most primal business, in which we invent and reinvent ourselves all day long, incessantly thinking and feeling, talking ourselves into being.

Saying local names and reinventing our stories is an endless, non-stop search for ourselves.

Places also come to exist in our imaginations because of stories. Having a "sense of self" means possessing a set of stories about who we are and where and how we live. As I suggested in *The Weekender Effect*, the mountain West is a place of myth, legend and story. Kittredge tells us that the stories we tell are important because they remind us all to love ourselves, one another and the world. Alexis might argue that loving ourselves, one another and the world, and expressing that love, is what language is for.

But something more is happening now which even the gods feared. It is more than the threat of loss of words and what they represent of the world. It is the threat of loss of the world itself. It is not just individual death we face but the loss of what is meaningful about life, at least in terms of how we value it now. Hermes, of all the gods, was aware of how such loss could occur.

Even the gods have no power over the three fates, Clotho, Lachesis and Atropos. As Alexis explains, Clotho spins the thread of life. Lachesis draws out the length of the thread of life each being will have. It is Atropos who then cuts the thread that ends each being's time on Earth. But often the threads of life are intertwined. As our fates are now so entwined with the lives of all other beings, Atropos cannot cut one thread without cutting others. What results from this is the collateral damage of extinction, whether accidental or otherwise. What we are witnessing in our time is the snipping apart of the love that connects us to a world we are rapidly losing. While this is happening all around us, each of us is on an endless quest to know what that love means before it is gone from the world.

Reading this book confronted me with the fact that all I really know about love is that it is unbearable to be without that which you love, and that it is that separation and loss that is the foundation of my own deep grief. I have been wrestling with this for a long time. You and your encouragement to read *Fifteen Dogs* demanded I go back ten years to once again face failures of courage in my life that Terry Tempest Williams challenged me through her writing to examine.

Williams argues we can save where we live, but only if we love it and are not afraid to act on that love. Through her work she reminds me constantly that "it is a vulnerable enterprise to feel deeply" and that we may not survive our affections. Williams feels, as I do, that we are being taught to hoard our spirit so that when a landscape we care about is lost – when, say, the town we live in is overwhelmed by thoughtless outsiders or eroded from within by parasites that feed on qualities of community and place; when, in other words, the places that define our identities are corrupted or taken from us – our hearts are not broken by this because we never risked giving our love away.

Williams is right. The onslaught of public relations and aggressive self-interest in our society has made us fear and suspect our deepest feelings of connection within us. By bottling up our cravings and our love and confining them within, Williams writes, paraphrasing Audre Lorde, we keep ourselves "docile and loyal and obedient" and we settle for or accept the inevitability of loss of what is at the root of our connection with place. And that, to a very real extent, is what I have done.

But Terry Tempest Williams won't accept my moral lapse, and worse than that she won't leave me alone. She believes in spiritual resistance – "the ability to stand firm at the center of our convictions when everything around us asks us to concede, that our capacity to face the harsh measures of a life comes from the deep quiet of listening to the land, the river, the rocks." It was because I believed her, and could no longer bottle up my love for where I lived and confine my rage within, that I wrote *The Weekender Effect*. But while the book has been read around the world, my protesting the loss of place and all that place means has had no effect. Instead of sounding an alarm, the book became a chronicle of the coming to pass of all that I feared. If anything, the onslaught of greed, self-interest and moral failure on the part of our leaders – municipal, provincial and federal – grew. I lost faith in my capacity to preserve meaningful qualities of the place in which I lived and once more began keeping everything bottled up. But once again, Terry Tempest Williams would have none of it.

In *When Women Were Birds* she called me out for my loss of moral courage: "To be numb to the world is another form of suicide." I realize now that I have trended and continue to trend self-destructively toward that end. My fear of loss has now become so great that it literally hurts me to visit the places I so care about. What I care about at the Columbia Icefield is literally melting away right before my very eyes. I am reluctant to drive downtown in the town in which I have lived for 35 years. It hurts to be reminded of so much loss. It is as if I were suffering from a form of "pre-traumatic" stress disorder – if such a condition could be said to exist. It is almost as if my heart is already broken, so there is no longer any point in risking giving my love away.

Even though I know it is unproductive to live like this, I have yet to find the strength to stand fully by my principles. Because I know that if I did I would alienate almost everyone around me, I just pretend to go along, with the crassness that is undermining my community, the destruction of place, the trashing of our national parks, the changing of the Earth's atmosphere, the acidification of the global ocean... as you know, the list of what we accept goes on and on.

But I also know – and we talked about this – that despair is pointless. I agree completely with you that there is no room and no time for despair. Despair, in fact, is the dead end of all dead ends. It is a cop-out. But I don't want to fall into the trap of blind hope, either, for hope without concerted action is little more than wishful thinking. But just as I feel that all I want to do is pull the covers over my head and pretend everything will take care of itself, here comes Terry Tempest Williams again, relentlessly poking the stick of reason into my protective moral cage. In *Refuge* Williams provides reasons enough to carry on with courage, not just in this life but in many lives hopefully to come. "The eyes of the future," she wrote, "are looking back at us and they are praying for us to see beyond our own time. They are kneeling with hands clasped that we might act with restraint; that we might leave room for the life that is destined to come." Then she nailed me right in the middle of the forehead. "Grief," she wrote, "dares us to love once more." Even though I have fallen off many times, I have to get back on the horse again if my life is to have meaning.

In *Fifteen Dogs*, Hermes – the greatest of all interpreters – was impressed with Prince – the last of the dogs still standing after they had been imbued with human ethos brought into existence through the understanding of language. To know, as Prince did, that his work and his language would with his passing disappear from the face of the Earth moved Hermes. That Prince understood that the death of language meant the death of its poetry impressed the god. Hermes recognized that to find oneself in that absence – to be essentially blind for want of something meaningful to see and deaf for want of something relevant to hear – would be unbearable. But, as you have pointed out, it is Prince's reactions to these circumstances that form the moral of the story. He chose joy. It was not easy but Prince reframed everything that happened to him around joy.

To think that something so essential as the manner in which we describe the world would die as a consequence of the world changing so rapidly and completely beyond what language is currently capable of describing is almost unimaginable. But that is what is happening now right in front of us. I understand completely the sudden surge of urgency at the end that compelled Prince to pass along the descriptions, the words, the connections and the poetry that was his world to anyone who might listen. He wanted dogs and people to love themselves, to love each other and the world.

For me, though, the question remains: Is this enough? Is the possibility of a seed sown, the hope of a reflowering of the world sometime in the distant future, enough reason for a committed life? Here I am, like so many of us, on a slippery slope, fighting beyond hope to prevent the devaluation of words and the cascading effects of that devaluation on the life-critical things those words describe, so that we can continue to justify saving them. I do not want to lose what gives our lives meaning and value in the face of ever-present death. But most importantly, I don't want to commit suicide by degrees by becoming increasingly numb to the world.

There is no need, Terry Tempest Williams assures us, to become numb to the world. Loss in the world is not always inevitable. Surprise is possible. Upon discovering a painted bunting that had stayed to winter in a New Hampshire town after being blown off course by a storm over the Atlantic, Williams compared the bird's situation to her own:

> Of course. Off course,... I have been seized in a storm of my own making. Whirlwind. World-wind. Distracted and displaced. In the wounding of becoming lost, I can correct myself. We can take flight from our

lives in a form other than denial, and return to our authentic selves through the art of retreat. ... And so we embrace the surprise.

The world can surprise us. We can surprise ourselves. We need to embrace retreat so that we can embrace surprise.

Terry Tempest Williams spoke for me when she wrote:

> I want to feel both the beauty and the pain of the age we are living in. I want to survive my life without becoming numb. I want to speak and comprehend words of wounding without having these words become the landscape where I dwell.

At the close of *When Women Were Birds*, Williams provides another avenue to the conclusion Prince arrived at in *Fifteen Dogs*:

> Once upon a time, when women were birds, there was the simple understanding that to sing at dawn and to sing at dusk was to heal the world through joy. The birds still remember what we have forgotten, that the world is meant to be celebrated.

Our challenge *really is* to heal the world through joy.

Williams is right also about writing as a potential means of redemption: "Writing is daring to feel what nurtures and breaks our hearts. Bearing witness is its own form of advocacy. It is a dance with pain and beauty." I am ready still to bear witness. I am once again up for that dance. Even though it breaks my heart to see it, I am afraid that given the rate at which Earth System decline is accelerating, I may have no choice but to transform my own nature, no other option but to translate my grief – as Prince did – into joy. It will not be easy, but if you do, I will too.

I am prepared to have my heart broken again, because heart is all I have. Thank you for recommending this remarkable book.

Your grateful friend,
Bob Sandford

Rivers Within Yearn for Rivers Without

Though water tends to repel organic compounds,
it is strangely attracted to most inorganic substances,
including itself.

Water likes to be around other water.

Its molecules, in fact,
cling to one another
more tenaciously
than those of many metals.

You can observe water's remarkable qualities
of self-adhesion
if you sit by a river.

Water sticks together.

Water draws water with it.
Sit on a riverbank long enough
and you might observe
that water likes to sing.

The faster it moves the louder it sings.

Still water barely whispers, falling water roars.

There is a reason we feel different

when we are in the presence
of large volumes of water.

Water reacts to almost everything
and almost everything reacts to water.

The feeling you get
standing on the edge of river or a lake
or beneath a thundering waterfall
may be aesthetic
but it is physical, too.

Your body is aligning itself
with the molecular attraction of the water
and the water is aligning itself to you.

The effect can be even more pronounced
when you stand by the sea.

Ankle-deep in surf,
the water in our inner cellular seas
yearns for the salty sea without.

The water within us feels the tug of the tide.
We know water, but water also knows us.

Go to the stream.
Watch the water run.
Feel the cool moisture of wind
and the wetness of cloud and rain.

Feel the cold of snow,
and the hardness of glacier ice.

Hear thunder.

Feel the river flow through your hands.
Feel the water within you
yearn for the water without.

Bring the stream to your lips.
Search with your tongue
for water's memory of faraway seas.

Taste distant mountains.

Feel the fissures in deep limestone tingle on your tongue.

Fill a glass and hold it up to sunlight.

See our star burn
through the sparkling lens
of the most amazing of all liquids.

Drink.

Repeat daily
until you
and the world
are fully and finally restored.

— R.W. Sandford, from *Water and Our Way of Life*

I am finally finding words for what I have seen and what I need to learn, not just about water but about life. But what I am finding is that there is something increasingly missing: a reason to carry on. Amidst so much diminishment and loss, all the old formulae for energizing my life and my work do not appear to work as they did in the past. Let's start with the increasing failure of the great writers of my time to continue to inspire me. I recently read a collection of what are effectively late-life essays by Annie Dillard, who has been an icon of example for me. I admired her understanding of natural history and her capacity for marrying science with writing that is not only transparent in revealing what science means but enters the domain of enduring literature in terms of the larger sensibilities it articulates. When I was 30, just reading the passage below would set me on fire for weeks. Certainly, it set me free when I first read it; but I am now having trouble keeping the fire lit. I can't get past the notion that I have somehow died, and not the fire. Here is what Dillard wrote:

> Why are we reading, if not in the hope of beauty laid bare, life heightened, and its deepest mystery probed? Can the writer isolate and vivify all in experience that most deeply engages our intellects and our hearts? Can the writer renew our hope for literary forms? Why are we reading if not in hope that the writer will magnify and dramatize our days, will illuminate and inspire us with wisdom, courage, and the possibility of meaningfulness, and will impress upon our minds the deepest mysteries, so that we may feel again their majesty and power? What do we ever know that is higher than the power which, from time to time, seizes our lives and reveals us startlingly to ourselves as creatures set down here bewildered? Why does death so catch us by surprise, and why love? We still and always want waking. We should amass half-dressed in long lines like tribesmen and shake gourds at one another, to wake up; instead we watch television and miss the show.

Is what is happening in the world killing poetry or is it just killing me? Dillard again:

> At its best, the sensation of writing is that of any unmerited grace. It is handed to you, but only if you look

for it. You search, you break your heart, your back, your brain, and then – and only then – it is handed to you. From the corner of your eye you see motion. Something is moving through the air and headed your way. It is a parcel bound in ribbons and bows; it has two white wings. It flies directly at you; you can read your name on it. If it were a baseball, you would hit it out of the park. It is that one pitch in a thousand you see in slow motion; its wings beat slowly as a hawk's.

One line in a sonnet, the poet said – only one line in 14, but thank God for that one line – drops from the ceiling.

I have worked hard and have felt the sensation of writing that Annie Dillard described. But I still can't quite find the words for the dread and sadness I feel as I am forced to stand by and watch the diminishment and loss of so much of the world I used to know. It is as if I am standing in the middle of slow-motion disaster and can't find or form words of warning that make any sense to those around me. Dillard notes that any culture tells you how to live your one and only life, and what you are told is that you must live it as everyone else does. But because of what I am seeing and feeling, I can't do that. Instead I feel paralyzed. If recall a famous passage from Dillard's *Pilgrim at Tinker Creek* that at first bowled me over, then inspired me greatly, when I first read it 37 years ago:

Thomas Merton wrote, "There is always a temptation to diddle around in the contemplative life, making itsy-bitsy statues." There is always an enormous temptation in all of life to diddle around making itsy-bitsy friends and meals and journeys for itsy-bitsy years on end. It is so self-conscious, so apparently moral, simply to step aside from the gaps where the creeks and winds pour down, saying I never merited this grace, quite rightly, and then to sulk along the rest of your days on the edge of rage.

I won't have it. The world is wilder than that in all directions, more dangerous and bitter, more extravagant and bright. We are making hay when we should be making whoopee; we are raising tomatoes when we should be raising Cain, or Lazarus.

Reinvention is about finding my way back to that inspired and purposeful way of living. How do I get out of this rut and start being truly alive again? How do I get myself out of the trap I gradually created for myself by slowly accepting the inevitability of diminishment and loss of the world? How does one reinvent oneself so that one can stand again in "the gaps where the creeks and winds pour down"? How can I prevent myself from sulking along for the rest of my days on the edge of rage? How can I instead start making whoopee again and raising Cain? And Lazarus. Right now I am like the monk Dillard described in her essay "The Waters of Separation," who carries his vision of vastness and might around in his tunic like a live coal which neither burns him nor warms him, but with which he will not part. I need that coal to warm me again. Though stalled, the reinvention, difficult as it is, must continue. I do it now or I am finished.

In *On Writing: A Memoir of the Craft*, the enormously popular horror-thriller writer Stephen King explains the honesty required to be a good writer – and you will note that I said *good* writer, as opposed to financially successful writer. To achieve any significant level of real competence, King advises, it is critical for a writer to be authentic to their own perception of what is true and meaningful in their own life experience. King committed early in his life to this dictum.

I was surprised, given his enormous success, that King and I share a lower-middle-class background. Here is how he describes his commitment to what interested him – not just what interested the people around him but what started *his* engine:

> And when I lay in bed at night under my eave, listening to the wind in the trees or the rats in the attic, it was not Debbie Reynolds as Tammy or Sandra Dee as Gidget that I dreamed of, but Yvette Vickers from *Attack of the Giant Leeches* or Luana Anders from *Dementia 13*. Never mind sweet; never mind uplifting; never mind Snow White and the Seven Goddam Dwarfs. At thirteen I wanted monsters that ate whole cities, radioactive corpses that came out of the ocean and ate surfers, and girls in black bras who looked like trailer trash.

I only wish I was as clear on my own interests at the age of 13 as King was, and as willing to defend and pursue them.

Among many other insights, King also offers a glimpse into his deepest interests as a writer – his obsessions, so to speak – which include how difficult it is to close Pandora's "techno-box" once it's opened. Anyone working in water and climate science today will certainly understand that theme. Exposure to King's deep interests made me contemplate my own. As King explains, he also wrote for the sheer joy of it. There were times also, he admits, when writing was little more than a small act of faith – a spit, as he described it, in the eye of despair. And then he offered a most important observation which for me is the single abiding reason I continue to write. Writing is not life, King observes, but sometimes it can be a way back to life. That is what writing is for me: a way back to life. A way of finding out who I am, a way of coming back to the potential and energy of life and finding the words for the beauty, joy, grief, sorrow and hope we need to carry on.

There is much in Stephen King's book that is both amusing and helpful to the writer, but nothing there was more valuable to me than his final few paragraphs, which I felt he in some quite unintentional way wrote for me at a time I was wondering if I still had anything left in me as a writer:

> Writing isn't about making money, getting dates, getting laid, or making friends. In the end, it's about enriching the lives of those who will read your work, and enriching your own life, as well. It's about getting up, getting well, and getting over. Getting happy, okay? Getting happy. Some of this book – perhaps too much – has been about how I learned to do it. Much of it has been about how you can do it better. The rest of it – and perhaps the best of it – is a permission slip: you can, you should, and if you are brave enough to start, *you will*. Writing is magic, as much the water of life as any other creative art. The water is free. So drink.
>
> Drink and be filled up.

As a writer concerned with water, I could hardly have been touched more deeply.

FIVE

THE HEART OF DRYNESS

> An "international water expert" grilling the old
> Bushmen about how humans must manage water
> was like a Vatican cleric interrogating Galileo
> about how the sun must orbit the earth.
>
> — James Workman

After arriving by way of Vienna it was good to be back in Africa.
A day after we arrived a dour but very polite guide drove us
around the savanna, where we observed the almost unimaginable variety of gazelles, impalas, zebras, hartebeests and hippos
that occupied the tiny reserve. We watched the snake-killing
secretary bird strut through the scrub looking, looking. We
came across an old male lion dozing almost invisible in the
tawny grass. We saw a black mamba slither through the grass
and then a green snake on the river where we were watching
for hippos bobbing in the slow, brown waters. We saw as many
animals as we had seen paintings in Vienna, and I was struck
by the potential relationship between the two. Could not each
of these animal sightings – carpeted with dozens of species of
grass, with the iconic flat-topped acacias in the distance beneath these exotic skies – be considered an artistic moment?

Certainly each sighting is a creative serendipity, a juxtaposition of natural events culminating in a continuously unfolding artistic here and now. As the pale and brief equatorial dusk fell around us, the sky brightened in the east and a gigantic orange moon rose alien in the African night. The lions would be hunting under its ghostly light.

The next day, I found myself hungry for meaning. The birds and animals were everywhere around us. How can it be that atomic elements, compounds and larger chemical complexes can organize themselves in such sophisticated and diverse ways? How can a single helix of DNA initiate such a grand procession of organic materialization? All around us there is physical evidence of the infinite numbers of consequent levels of interaction that emerge from elemental combinations. Single genotypes are the nuclei around which nerves, organs, muscles and brains organize themselves. The body is a locomotive transporting the fantastical idea of a rhinoceros through the grasses and the trees, themselves simply phenotypical materializations of still other gene constellations around which life forms. The hunger of the incomplete atomic subshell is never satisfied. Instead this hunger is projected for all eternity onto ever larger screens: eager viruses, aggressive bacteria, lonely sponges, predatory fish, patient reptiles, trumpeting elephants, thorny trees, puzzled people, attracted planets, ecstatic galaxies – an entire universe hungry for wholeness. But fundamental wholeness was made impossible at the moment of creation, at the Big Bang, when an imbalance in the composition and distribution of the elemental constituents of the universe decreed that all of the stuff of being would remain forever poised to become something else: soil aching to become a plant, plants always on the cusp of becoming animals, animals in the constant terrible tension of simply being, before their own chemistry dissolves them and the hunger that is life rebuilds them in a new form. So we will always have diseases; lions will always attend the herds of wildebeest; there will always be warring cultures; and there will always be thirst.

It is hard to imagine how dry the central part of Tanzania can be. Tall dust devils rose eerily from the desert-like plains amidst which Masai herdsmen could be seen tending flocks of skinny goats or droves of Brahman cattle. What these animals ate and drank I have no idea. It seemed the soil was only fine dust, and that much of it was being carried up into the little tornadoes that rose in brown spirals a dozen at a time from the heat-light of blistering, naked earth. We pulled into a truck stop for fuel and oil for the rattling Nissan van. Immediately we were surrounded by milky-eyed Masai women begging for pens, soap and money while trying to sell us bead necklaces and earrings. I observed these gentle people for some time. I wondered what would become of them, the soil blowing away, the earth scorched by a million years of sun, the country over-crowded, poorly governed, with only a meagre hope of tourism to bring in badly needed currency to augment diminishing for-eign aid. No wonder they crowded around the van. What can you say about the people who live in this ocean of fine dust? They are beautiful and kind and they do make lovely baskets, but they face an uncertain, thirsty future.

Later, coming down a hill from the tourist lodge to the flat plains that are now protected as a national park, we learned from another driver that poachers had shot two wildebeest in-side the park the night before. It seemed that everything was breaking down. I was reminded of a book of essays I had read before I left Canada about life in contemporary Botswana.

Whites, by Norman Rush, gives the reader a frank, tough view of what it was like living in post-colonial Africa. Rush makes an interesting observation early in the book about how the climate and nature of the continent seem to breed breakdown even as a function of its natural processes and cycles. In the story "Near Pala," one of the characters remarks that during an especially long drought, thirsty animals began fighting over carrion. He indicates that a natural protocol between vultures and jackals, for example, was breaking down. In ethological terms, jackals typically withdraw from a carcass when the birds come, but during the drought they were staying and fighting. The same breakdown in natural protocol is exemplified at the climax of the story, when the characters driving over a dusty road are hailed by a small group of native women begging for water. The women are emaciated from hunger and thirst and dressed in rags and skins. With unnatural smiles, they are calling "Metse, metse," their word for water. Despite the pleading of his wife, the driver of the Land Rover will not help the thirsty Basarwa, claiming the vehicle would get stuck in the sand if they stopped. A woman runs beside them pleading, but they didn't slow down. She eventually falls to the ground with exhaustion while the vehicle continues on in the dust.

When it comes to thirst, the dry parts of Africa are a bellwether for the rest of the world. The situation in Africa and its implications for the rest of the world is carefully described by James Workman in his 2009 book *Heart of Dryness*. This is a rare and important work in that it gives meaning and value to recent events to which most North Americans would have paid little if any attention: the removal over the previous decade of the Gwi and Gana Bushmen from the Central Kalahari Game Reserve in Botswana. What makes the book so valuable is that

Workman was able to derive considerable symbolism from the suffering that took place there and to use that symbolism to demonstrate how our inefficient management practices are affecting those who derive their livelihoods and identity from a close understanding of arid ecosystems.

Workman begins by pointing out that there are as many ways to run out of water as there are ways to define drought. This suggests that the ways we can create water scarcity through our own actions are likely as numerous as the ways nature itself creates scarcity, which, as it happens, is a lot of ways. Workman is one of the first popular writers since the climate change debate got going in earnest to take up the issue of permanent drought, not just in the usual places like Africa, the Middle East or Australia but in North America as well. Workman understands drought on a global level. Far from representing a universal absolute, Workman tells us, drought typically describes a subjective and relative condition that varies markedly by population and place. Set against the background equilibrium patterns of previous decades or centuries, measurable drought occurs whenever mean temperatures escalate for a prolonged period, causing water tables to sink deeper, evaporation rates to increase, reservoirs to drop. In such circumstances dry seasons last longer, and the economic thirst of more people demands more water for more purposes than before. Nothing prevents all of these unpleasant phenomena from appearing at once and thereby compounding their effects. As Workman points out, many scientists believe that exactly this kind of convergence is happening just now in North America, a fact borne out during the ten years following the publication of his book in 2009.

Experts in water resources, according to Workman, had identified what was already happening at the time, and is continuing today in the United States, as a perfect perpetual or permanent drought. This means that in spite of unprecedented prosperity and freedom in other sectors, and the billions of dollars worth of bottled water on supermarket shelves, Americans enjoy less absolute access to water than ever before. Workman points out that this is a problem less connected to supply than to demand. When Europeans discovered the New World there were perhaps 14 million people living on the North American continent, each using somewhere around 500 or 600 litres of water a year. Now each of some 300 million Americans and 34 million Canadians use that much each day, plus an additional 20,000 litres a year for food and in the manufacture of the goods and services to which they have become accustomed in their daily lives. Water demand rises dramatically with affluence. As our houses become bigger and we eat better, our use of water grows. Most North Americans find themselves crowded into bigger and bigger cities highly dependent on deteriorating water infrastructure that was designed for significantly lower needs in historically wetter eras. As a consequence of population growth and pollution, which limit the number of uses to which water can be put, each North American has available less than half the water the previous generation had. In addition to this, our species now takes so much water for its own use that aquatic biodiversity is now disappearing faster than anywhere else in nature.

In addition, because of landscape and climate changes, we find ourselves competing for less and less total water even as our per capita availability shrinks. Many formerly cool, wet places are drying up. Abnormally arid conditions and heavy human demands for water have not only dried up the Colorado as it heads for the Pacific and the Rio Grande on its way to the Gulf of Mexico but also desiccated all or part of 49 states, shrinking 57 other American rivers to record low flows. Historically, when dry conditions plagued states in the midwestern or western United States, people simply moved to more water-abundant places. That is no longer possible, however, as 36 U.S. states are projected to face persistent water shortages within a decade.

In the past, Indigenous peoples stockpiled food and stored water while waiting for cyclical dry spells to end. That strategy may not work much longer, as our collective water demands have become too great. More troubling is the increasingly obvious fact that the wet period during which Europeans settled North America, and around which our water infrastructure needs were defined, could come to an end at any time. In this context the picture is frightening enough even if we don't take climate change into account.

Droughts in the historical record over the last millennium were deep and persistent enough to give us an idea of the kind of future we are going to face even if our collective human activities weren't poised to cause our climate to warm. The people who lived on this continent before us endured droughts that lasted decades. We know also that despite the great adaptability of these earlier peoples, many of them did not survive the extended periods of aridity that shrank water supplies and reduced the availability of food.

Some scientists are saying we may be in for the driest period in North America in 30,000 years, and that for all intents and purposes the drought we face may in fact be permanent. In the past we could rely on trade with reliably water-abundant European countries, but now even they are facing limits to how much food they can grow in an increasingly water-scarce world.

The same thing has begun to happen in the rest of the world. Reliable precipitation is becoming more uncommon and food production more uncertain.

Water scarcity is no longer just an agricultural or environmental concern. It is now seen as a potentially limiting factor in the further economic development of many parts of the world. By doubling every 20 years, world demand for water is seen as unsustainable. In a recent survey, two of every five global industries declared that the impact of water shortages on their business would be severe or catastrophic. Seventeen per cent declared they were not prepared for such a calamity. The most serious problems generated by water scarcity, however, may ultimately be political. Drought, Workman points out, tends to divide people at the individual and subnational level. While countless energy initiatives proclaim it will be possible to go "beyond petroleum" as an energy source, there is no substitute for water.

Workman points out that the problem is also institutional, in that we have discovered we can't "regulate" the decreasing flows of rivers and declining aquifers any more than we can "regulate" climate, clouds or rain. In Workman's estimation, the Bushmen of the Central Kalahari have got it right. They continue to survive in one of the driest places on Earth because they accept that we do not govern water. Water governs us.

Workman asserts there are not two but three inevitabilities in life today. Besides death and taxes there is heat. Rising dry heat is sucking moisture from soils all over the world. Drylands now comprise some 40 per cent of the Earth's land surface, and a third of the world's population lives there. As an international expert on water resources, Workman has felt the warm breath that is melting away the Earth's snow and ice, and that has inhaled anemic rivers and evaporated drinking water reservoirs in major cities like Sydney, Mexico City, Jerusalem, Beijing, Tehran and New Delhi.

Observing that his crowded planet had been diagnosed with "a terminally warmer future," Workman did the one thing he thought would teach him what he needed to know in order to live in that future: he sought out the most experienced survivors of perpetual drought and water scarcity in the world and went to live with them in the Kalahari Desert in Africa. It was there he learned how people living on the land actually survived droughts of 50 years in duration.

Workman did a great deal of research before he went to live with Bushmen in the last defence of their Kalahari home. He learned that in tracing the earliest human roots, geneticists had discovered a point in human history when a fledgling *Homo sapiens* nearly went extinct. The cause was a cataclysmic drought that swept across all of Africa, wiping out all but a few thousand people. Paleontologists now also hold that it was yet another megadrought that scattered many of our human ancestors all over the globe. Only the hardiest of the survivors remained in small, isolated pockets on the continent of Africa. Among these were the ancestors of today's Kalahari Bushmen.

Workman describes how researchers traced DNA samples back 150,000 years until a single female bloodline emerged from a primordial Eden. It is from this bloodline – and the anthropological Eve that began it – that all humanity has descended. Workman notes that the longest genetic strain passed down from the earliest drought survivors is today collectively shared by the bands of Bushmen who live in the Kalahari Desert. The point he wants to make is that among the elder matriarchs of this people there are living individuals whose genetic makeup is a modern reflection of Eve, the first woman, to whom the human world owes its origins.

Workman also notes that, despite their incredible genetic lineage, these proto-people are not respected. The reason they came to his attention at all was that the land where they lived had been expropriated by the government of Botswana for a park. The government had cut off water supplies to Bushmen villages as a means of forcing the remaining inhabitants into agricultural communities and urban enclaves. Some Bushmen, however, resisted and continued to live in the desert, surviving only on the limited supplies of water they could find there. Workman's book is their story, but it is also our story. Two thousand generations after humans emerged in Africa, the 7.6 billion people currently living on Earth are no less vulnerable to drought than our ancient ancestors were.

The Sahara is advancing southward at a rate of 50 kilometres a year. Vast regions of Africa's degraded and rapidly expanding Sahel region are drying out, and once-productive soils are being scooped up by the wind and deposited as useless dust in southern Europe. Workman cites a research report that says at least half of the dust in the air today has come from Africa. But other places are drying out too. Atmospheric dust has increased by a third over the last century. We know that surface moisture moderates climate; and the hotter it gets, the less surface moisture is available to moderate climate.

Workman fears that as we continue to burn and clear forests, convert land to irrigation agriculture, and power an entire society with fossil fuels, we are unwittingly baking the Earth in what may well be an irreversible progression, at least in terms of meaningful human time frames. Our carbon emissions, he observes, have thickened thin layers of the outer atmosphere, trapping solar radiation. The effect, Workman warns, "resembles leaving our collective car in an exposed parking lot with windows sealed and the kids locked in."

This is a troubling analogy, for it suggests that in denying climate change and delaying action to counter the threat, political leaders are effectively decreeing that we as a society are just going to leave future generations to cope without hope.

Workman points to evidence that the jet stream in the northern hemisphere – the horizontal conveyor belt of cold air that dips down from Canada across the United States and creates national weather systems – has been migrating northward at a minimum of 20 kilometres per decade, or 5.5 metres per day. As it moves up through a region, high pressure and clear skies converge in its wake, which is leaving the U.S. South and Southwest hotter and drier. While many animals can advance north with the migrating jet stream, habitats cannot. Workman quotes Ken Caldeira, a climatologist with the Carnegie Institute, who urges Americans to look south of where they currently live and "that's probably a good idea of what your weather will be like in a few decades."

Both the World Meteorological Organization and the British Met Office confirm that the past decade was the hottest ever. For the latter seven years of it the United States broke records for high heat and low reservoirs. We appear to have entered a new hydrological regime. But who, Workman asks, can teach us how to cope with drier conditions? Meanwhile, the Kalahari Bushmen, the world's oldest surviving civilization – and the only people with the survival experience, strategies, tactics and values to guide us through the extremes of our once and future drought – are being driven out of their homeland by armed men urging the last free Bushmen to surrender their way of life forever. This, Workman suggests, is a highly symbolic situation.

He notes as well that some sieges are unintentional and produce no victors. Climate change impacts on water availability may end up being one of these. Our unstable atmosphere has already begun to alter rainfall patterns, reducing runoff and storage capacity and accelerating evaporation in drylands which, as mentioned earlier, comprise about 40 per cent of the world's land surface and a third of its population, spanning some 80 countries. Even in the unlikely event the world would act in a unified way to address the climate threat – and even if all emissions ceased immediately and no further carbon found its way into the sky – the world would still keep warming for the next hundred years. In the meantime, populations will keep growing, glaciers and ice caps will keep melting, snowpack and snow cover will continue to be affected and tropical belts of aridity will go on expanding, converting forests into grasslands and grasslands into deserts. Paraphrasing U.S. cartoon character Pogo, Workman says that in our conquest of nature, civilization has managed to besiege itself: "We have met the enemy and he is us."

Workman's book also explores how the dictatorship of drought shapes the existence of everything that lives in the Kalahari Desert. The camel thorn acacia sinks deep tap roots to extract every precious drop of water from pores between the sand. The baobab's spongy bark is saturated with water that the tree uses to endure prolonged drought. Other plants chase water in slow motion. Workman goes so far as to say that the prime catalyst behind natural selection in arid regions is not oxygen, sunlight, fire or soil; none of those was necessary to support life. Rather, the force that governs existence in drylands is the relentless quest for water. The reason why people who live permanently in the Kalahari remain inactive during the heat of the day has nothing to do with lassitude, sloth or surrender. These people habitually and deliberately reduce energy expenditure to conserve water for essential demands. This calculated restraint in the face of dry heat reveals the oldest, most energy-efficient and, from an evolutionary standpoint, most effective form of living within the rule of water.

According to Workman, we are about to relearn what it is like to live under the rule of water in North America. Of the 100 fastest-growing cities in the United States, 92 are in water-scarce areas. To make matters worse, in recent years up to a third of global water withdrawals were directed to irrigating new sources of fuel. Since it takes 9000 litres of water to produce one litre of biofuel, and 4000 litres to produce a litre of corn ethanol, a "clean fuel" drive to the local grocery store could burn up more water than would have otherwise been used to grow a week's worth of calories.

As Workman points out, there was a remarkable absence of planning in the drive to convert food into fuel. Either advocates didn't think about the amount of water biofuel production required, or they simply chose to ignore the fact. Either way, the resulting food-price inflation brought widespread misery so deep and pervasive that it defied statistics. What effectively happened in places like Mexico and elsewhere is that the middle class cut meat, the lower classes cut vegetables and the impoverished stopped feeding themselves and turned to food aid.

We have once again been reminded that food production isn't as dramatically limited by land, innovation, price, know-how, support, fertilizer or technology as it is by the availability of water. Without water all of these other factors don't matter much. Upon reaching this verdict, Workman continues, experts developed a multi-pronged solution to the problem of supplying enough food for our rapidly growing global human population. This remedy involves:

- squeezing more productivity from each available drop of water;
- devolution of authority over water to informed local decision-makers;
- diversification of crops beyond risky monocultures;
- reduction of the distance between where food is produced and where it is consumed;
- abandonment of inefficient and water-intensive agricultural practices;
- adaptation of crops to local rainfall regimes; and
- exploitation of "green water" in the ground.

According to Workman, this agenda looks good on paper, but it is not working as effectively as it should, because powerful agro-industrial interests feel threatened and rejected by it, as is certainly the case in dry Western Canada. Workman observes, however, that these strategies mirror exactly those employed by the Bushmen of the Kalahari, who have been living this "new paradigm" of hydrocentric agriculture for centuries.

Workman goes on to point out that at less than 23 inches (584 millimetres) of rain annually, wild ungulates out-compete domestic stock. More specifically, endemic species do better than domestic livestock in fighting disease, surviving predation and persisting in the face of drought. In addition, domestic livestock need more land than wild ungulates, which is why contemporary ranching is driving Indigenous peoples and the animals they hunt out of existence in many places in the world.

Another huge influence on desert life has been the emergence of boreholes. In Workman's estimation boreholes have done as much as anything in the Green Revolution to stave off famine, feed India's billions and save lives in Africa. In the Kalahari, 83 per cent of all rain evaporates and 14 per cent transpires through plants, leaving only 2 per cent to trickle into ephemeral pools and pans. But some 21,000 boreholes have been sunk to bring groundwater to the surface for the exclusive use of cattle ranches. And still this may only be a temporary solution. Tipping the scales at one ton per 265 gallons, water is heavier than oil. Lifting water from deep in the Earth requires continuous and growing reliance on expensive fossil fuels. The deeper the well, the costlier it is to draw water up it. Breakdowns and expenses often bring pumping to a halt. Thousands of Botswana's boreholes sit broken, rusted, clogged and corroded from neglect. On average, boreholes last 30 years, but in Botswana many no longer work after 10.

Workman observes that given enough cheap fuel, tapping invisible groundwater is like grocery shopping on someone else's credit card. Sooner or later the account will dry up. Botswana abstracted fossil water from its subsurface aquifers at more than 900 times the rate of recharge. As has happened in the Ogallala Aquifer in the United States, the decreasing volumes of groundwater remaining were too deep to economically pump, and wells began going dry.

The moment water becomes scarce, wealthy and powerful interests intervene to ensure that common needs do not affect the status quo. In theory the presence of so many boreholes in Botswana should have led to the decentralization of authority over water resources to informed local decision-making bodies. That did not happen. The central government owns and controls half of the country's boreholes. Select cattle barons control the rest. Even when well-meaning foreign charities drill for water to help marginal peoples like the Bushmen, their wells are quickly appropriated by politically connected ranchers. Some of these ranchers, Workman reports, create quasi-feudal fiefs around their control of water, granting access only in exchange for physical labour or sexual favours.

Ranching also inhales water in another, more insidious way. In 2002 the 1.5 billion cattle in the world ruminated 18 per cent of the world's greenhouse gases into existence, mostly in the form of methane and nitrous oxide. Methane is 23 times more potent a greenhouse gas than carbon dioxide, while nitrous oxide produces 296 times the greenhouse impact of CO_2. As Workman demonstrates, cattle generate a greater greenhouse effect than all the cars and trucks in the world combined. He quotes Bryan Walsh, a contributor to a 2007 *Time* magazine cover story, to the effect that given the energy consumed in raising, shipping and selling livestock, "a 16-ounce T-bone is like a Hummer on a plate." This view doesn't appear to be having any effect on the popularity of beef, however. Beef production is expected to double to 455 million tons by 2050. This means that despite knowing better, governments will continue to eradicate diverse water-adapted game species in order to install thirsty European livestock on dry landscapes, which in turn will require the further displacement if not slow-motion genocide of intransigent Indigenous locals. This will ultimately extinguish the age of the hunter – and the age of true adaptability to drought acquired over hundreds of generations of human experience in the dry times of the past. This was exactly what Workman witnessed happening in Botswana.

With rising temperatures we will also have fewer and fewer

places to find respite from heat and drought. Some of our established water management solutions will no longer work. Again in Botswana, for example, Western experts instructed the government to build large dams on thin rivers and keep the reservoirs full for as long as possible. But when this was done it was discovered that evaporation rates were so high that most of the collected water was disappearing into thin air. In order to evaporation-proof the country's water supply, experts recommended artificial aquifer recharge. It was soon discovered that forcing water into aquifer fissures and pores saved four out of every ten gallons of evaporation, allowing flexibility and security in water supply. Unfortunately, however, an analysis of the best potential underground storage aquifer found it contaminated with nitrate pollution from agriculture, making effective artificial recharge difficult if not impossible. As reservoirs in the country continued to decline due to deep and persistent drought, the government of Botswana realized they had done what no Bushman would ever do, and that was to sacrifice and foul the very waters the country would need to survive. It was soon realized – as it has been everywhere else in the world where this has happened – that once you contaminate groundwater, it is difficult if not impossible remediate the damage. Even if you could afford to fix groundwater pollution, it is beyond our technological capacity to do so within any meaningful human time frame.

Instead of repairing leaky water infrastructure – which was losing three litres of water for every litre used – the government spent $7-million on new pumps which forced water at a higher pressure through a broken system so that water losses were even higher. Pressure in the water distribution systems would build up during the night and be forced out through cracks, sometimes doubling or tripling the leakage. Squatters and the poor were also tapping into water mains illegally. The fact that the government subsidized water use created perverse incentives to ignore the problem. In effect the government was the biggest water client, and it was also the biggest water waster.

Workman points out that what happened in Botswana is hardly unique. In nearly all big cities – from Los Angeles to New Delhi to São Paulo and Mexico City – at the time he wrote his book, 40 to 60 per cent of treated water was never making it to the intended customer, due to leaks, corrosion, disincentives and poor infrastructure maintenance. Farmers worldwide, Workman reported, lose two-thirds of their water for similar reasons. In India half of the nation's water is lost between source and delivery because of leaky pipes. Yet farmers will not permit the introduction of water meters. China – now the planet's thirstiest country – wastes more water from aging infrastructure than it can store. Developed countries are not necessarily more efficient. Even during the recent drought, system leakage in the U.S. Southwest amounted to more than six billion gallons a day. And it will be costly to fix the problem. According to Workman, the unmet costs of water infrastructure repair in the United States alone are estimated to be somewhere between $500-billion and $800-billion.

When the Botswana drought persisted and water supplies dwindled, the government was forced to employ highly unpopular measures to reduce consumption. In the same way it had cut off water to the Kalahari Bushmen to push them out of their desert territory, the government began to stop or severely restrict water supply to own citizens. In the short term these measures worked. Within six months Botswana reported a 35 per cent drop in water use. But then the other shoe dropped. In rationing water, the government imposed sudden, inequitable and sometimes permanent social, economic and environmental devastation. Trust in government itself was shaken. Water-intensive companies were forced out of business. Workman reports that real estate prices plunged, undermining new development and economic growth. Ruling political parties lost support.

Even water utilities were undermining their own future. As rationing reduced revenues, utilities laid off their most skilled and experienced workers, leaving only the lowest-paid staff, who were less able to monitor reservoirs, investigate leaky pipes, explore coping strategies and enforce the rationing that had been imposed. Instead of creating austerity, water rationing simply generated incompetence, which soon threatened the capacity of the government to run the country. People turned on one another, informing on their neighbours if they violated water restrictions. Civil order began to fragment.

As the economy declined, crime rose, along with xenophobia. Some of Botswana's most affluent citizens and expatriates discovered that drought had turned the country into a less than desirable place to live. Meanwhile the heat island effect in the country's capital was making the city almost uninhabitable.

In order to quickly find out what it was still doing wrong, the Botswana government hosted an international water symposium attended by 90 experts from five continents whose countries could be facing the same situation as Botswana. The common threats were easy to identify. They included rocketing populations, rapidly growing water demand, shrinking supply, stresses from desertification, and effects of global warming. To the government's astonishment, Workman reports, the symposium concluded that the building of the Gaborone Dam – once the nation's pride – had been a grave and potentially fatal mistake. It appears that centralized, large-scale reservoirs make arid countries increasingly vulnerable to climate change. It was the consensus of the symposium participants that big, expensive water supply projects concentrated populations and increased dependence at a time when cities should be dispersing and decentralizing demand. Water storage dams only make sense in rainy countries, Workman asserts. In arid regions like Botswana – or for that matter the western United States, southern Europe and perhaps even the prairie regions of southern Canada – big dams actually increase the risk of drought. New dams could be built and even connected to aquifers for recharge, but this approach won't work if a warming climate results in consistently reduced streamflows.

With average temperatures rising by a degree every few decades, big dams will not help during the more frequent and persistent droughts many countries are projected to experience in almost all climate-warming scenarios. Far better than big dams, experts offered, were small-scale, non-centralized, locally managed, autonomous reservoirs where rain could be efficiently trapped and stored to then be carefully transported and productively used. This, Workman notes, is pretty much the way the Kalahari Bushmen have managed water for thousands of years. Unfortunately it is difficult to go back to this way of handling water resources if water has already been made into a commodity whereby profits are made through market trades, or if water is no longer held to be a human right. Workman reported that private water services companies worldwide were expected to do a trillion dollars a year in business by 2015, and they did. As the Bushmen were to discover, issues related to water as a human right remain unresolved.

It is held by many that it ought to be self-evident that water would be a fundamental human right. The argument made by Maude Barlow and others in Canada is that, given that water is the essential element for life, it must be secured for the people by the government against the interests of business. This argument hinges on the notion that fresh water belongs to the Earth and to all species, and therefore must not be treated as a private commodity to be bought, sold or traded for profit. The global freshwater supply is a shared legacy, a public trust and a fundamental human right.

The opposing, "conservative" argument is that the trouble with declaring things like water and sanitation to be human rights is that it often achieves the exact opposite of its aims because it invites state intervention into a broad range of domains in which government has no place. At the extreme, opponents of the notion of water as a human right argue that instead of conserving water for all species, such a declared right would quickly lead a nation to waste, pollution, corruption, biodiversity extinctions and state insolvency. As Workman says, a lot of people can fall through the cracks between these two opposing views.

Workman further notes that freedom from thirst is not mentioned in the UN Universal Declaration of Human Rights. The country that has done the most to make sure it never makes that list is the United States, and at present Canada supports the U.S. view. United Nations platitudes aside, there is no legally binding global right to water. Workman nails what it will take to make water a human right:

> All rights – and limitations on the state – emerged through ugly and messy processes, repeatedly and

violently clawed and scraped and forced into the light
where they could be defended.

And in fact, the relationship between drought and violence
is becoming clearer. While nations might not quarrel directly
over water, they will battle over the absence of it. In western
Sudan, for example, fighting broke out in the 1980s and '90s between farmers and herders after the annual rains failed. In the
ensuing genocide more than 200,000 people died and millions
more were forced to flee their homes.

While Iraq, Syria and Turkey might not wage war directly
over the transboundary flows of the Tigris and Euphrates, all
three have systematically killed Kurds who live in the region's
upland watersheds. It is now becoming more common, as we
see in Israel, for sovereign nations to deny stateless people adequate water. Israel made peace with Jordan but occupied the
Palestinian West Bank in part to secure its three vital aquifers.
China did not fight Burma or India but did absorb defenceless
Tibet, the upland source of two-thirds of Asia's water. Stripped
of their rights and deprived of water resources, usually unarmed
dissidents have nothing to lose, so they fight back. When faced
with such opposition, water-hungry governments taunt resist-
ers until they do something the government can kill them for.
When charged with genocide in their treatment of the Kalahari
Bushmen, Botswana hired a multinational public relations firm
to make villains out of the weaponless people they killed. With
the help of such propaganda the government turned the Bushmen into terrorists. And terrorists they apparently remain.

What distinguishes James Workman's book from so many
others on the topic is that he brings a great deal of relevant
supporting information together to give enormous emotional
power to his unforgettable conclusion. When Bushman elder

Qoroxloo Duxee, the heroine whose life he chronicles throughout the book, dies of thirst, Workman carefully details the results of the autopsy to come to a stunning conclusion which
nests Qoroxloo's story and that of the Kalahari Bushmen into
the larger global and symbolic reality of what drought may
mean to us in a hotter future world.

Workman sets up his powerful conclusion by detailing
exactly what happens to the human body when a person dies of
thirst. As water grows scarce, the organism begins to compete
with itself. Cells vie with one another over the common water
they all need. Voluntary muscle tissue competes with involuntary tissues. Blood, consisting as it does mostly of water, has
its own needs, as does the heart. Stripped of its romance and
mysticism, the heart is really just a pump. In the average human it beats perhaps a billion times in its role of replenishing
the body with enriched water to satisfy the thirst of the organs,
including the heart itself. Each organ claims supremacy of need.
This is particularly so of the tissues of the nervous system.

The brain, composed as it is of 90 per cent water, competes
as hard as the heart for water. When examined, Qoroxloo's
heart was revealed to be atrophied and a discoloured brown.
The physician conducting the autopsy observed that Qoroxloo's
heart showed clear signs of chronic or long-standing dry star-
vation and dehydration. To the doctor this was evidence that
Qoroxloo's heart had sacrificed itself over months or even years
in favour of sending more water to her brain. Those months and
years during which she was sacrificing her heart to her head to
keep her family together were the very months and years when
the government of Botswana was denying the Bushmen water
in order to force them out of their desert homeland.

For years, Qoroxloo denied herself the water her lifelong

experience had allowed her to find so that others could eat and drink – a sacrifice Workman calls the ultimate form of human adaptation. But even after offering such symbolic evidence of what it will take to survive in a hotter, more arid world, Workman is still not done with the reader. In a hot, dry world, he explains, water equals power. We may not like the rule of increasingly scarce water resources, but at the same time we cannot escape it. The Bushmen of the Kalahari have demonstrated for thousands of years how to embrace that reality. Their fundamental rule of adaptation is not to organize and mobilize physical resources to meet ever-expanding human wants, but to organize human behaviour around constraints imposed by diminishing physical resources. At present, in the Canadian West, we still have a long way to go to learn this lesson.

Ultimately, Workman reminds us, we do not govern water. Water governs us. In dry regions of the world, water scarcity defines the outcomes of all vital decisions. In the Kalahari, water scarcity determines who and what can be trusted and why; where and when to disperse; what to eat; how much to consume; and which plants can be burned for fuel and which used for construction or gathered for the water in them. A diet that is water-secure emphasizes diversified, nutritious, drought-resistant and moisture-rich permaculture over tastier, storable, transportable bulk foods. Since tastier feedlot cattle could not survive droughts, the wise hunt desert-adapted game species whose juicy meat concentrates metabolic water.

If the Bushmen ran our water supply in North America, Workman believes they would organize us around the measurable contours of the hydrological unit where we live: the water known to exist in an aquifer or river basin. Then, within that unit, their code would secure the fundamental and minimum amount of water required to keep each human alive and healthy. This volume they would declare a fundamental human right of the same kind the Bushmen held to be theirs in the Kalahari. The flip side of this right, however, would be responsibility for that water. The Bushmen code would likely then have them suggest that, given the finite limits imposed by drought, people would be allowed to negotiate informally over the water they require, reaching out to partners to trade with if and when they needed more or less. People would be expected to increase supply by efficiently reducing demands, and the benevolent result of this integrated right to water would bring society back into a relative state of social abundance.

Workman concludes that water should not belong to the government. It belongs to each of us, or would if we hadn't already given it away. He argues that public or private utilities are neither good nor evil, but right now they still remove all real incentive and accountability for conserving water efficiently, while making us depend on aging infrastructure, political fecklessness, wasteful management and unreliable supply in the face of a changing climate. In an era of permanent drought, this is not a good place to be.

Workman likes the idea of unlimited markets within natural monopolies. The best part of this approach, he argues, is the political benefits it would bestow. In our current circumstances, splinter interest groups pressure politicians to keep water rates low, build more dams, drain more wetlands, pump more deltas, expand storm drains and sewers and plunder more aquifers. If we nudged governments to raise rates across the board, however, we would reward efficiency, make the water we conserved more valuable, drive ourselves to more efficient exchanges and restore substantially more leftover wild water back to beleaguered and endangered aquatic species and ecosystems. Workman believes that by redefining water as an owned and tradable right, we can turn costly conflict into symbiotic co-operation that will alleviate national-security tensions over transboundary rivers and aquifers. Not everyone agrees, but Workman concludes his book brilliantly:

> When I think of the permanent drought we face in the years ahead, I like to picture Qoroxloo as last seen by her band of foragers: calm, defiant, and aware, striding purposefully across the hot, dry Kalahari sands while quietly singing an ancient song to herself … and to anyone else who might care to listen.

SIX
IRRIGATING EDEN

1. ISTANBUL

There ain't no doubt in no one's mind
That love's the finest thing around
Whisper something soft and kind
And hey, babe, the sky's on fire, I'm dyin'
Ain't I, gone to Carolina in my mind

— James Taylor, "Carolina in My Mind" (1969)

I first visited the Middle East when I was in my 20s. I had never been off the continent before and suddenly arrived in Egypt with my mother, who wanted to visit the Holy Land before she died. I was so overwhelmed that I kept singing lines from a James Taylor song just to remind myself that I had a culture I could cling to. The trip blew the vault of my mind open and I have never been able to close it again. The whole world entered into the rent which that trip made in my consciousness, and I have been travelling ever since. Travel is how I put myself into relief. It is a way of comparing where and how I live with how others define themselves. As Lawrence Durrell once wrote, travel is "an outward symbol of an inward march upon reality."

When I returned to this part of the world, decades later, it was to attend a conference of the Rosenberg International

Forum on Water Policy. This time, however, my voyage was more than just exploration. It was also an inquiry into the state of the planet as indicated by how we manage the world's fresh water.

I worked for a couple of hours in the lounge before finding Kindy Gosal, Josh Smienk, John Harrison and David Hill at gate 64 in the eternal city that is the Frankfurt airport. Our flight to Istanbul was a comfortable two and a half hours during which I discovered I was seated next to Dr. Reg Gomes, vice-president of the agriculture and natural resources department for all ten campuses of the University of California. The affable Gomes explained that he had been involved in the Rosenberg Forum from its inception in 1994, when, at retirement, Bank of America chairman Richard Rosenberg endowed the University of California with the resources it would need to offer a high-level biennial invitational global forum on avoiding conflict over transboundary water issues.

Just as Rosenberg co-chair Henry Vaux had indicated, the airport in Istanbul is spacious and beautifully modern. I was most pleased to find my luggage had arrived from Calgary. The ticket agent there had put a tag on my suitcase based on the first two letters of my destination, but hadn't checked what was after the I-S to see if those were right the coordinates for my ultimate destination. It was most fortunate indeed that I noticed she was sending my luggage to Islamabad instead of Istanbul.

When the van we were expecting from the hotel didn't materialize, we got into two taxis and convoyed from what at that time was called the Istanbul Atatürk International Airport to the coastal freeway and then on toward the old city. We were amazed by the grand views the route granted us of the Marmara Denizi, the Sea of Marmara. We watched fishermen with long poles cast their lines while traffic bustled by on the sea in front of them and on the freeway at their backs. The land and the water appeared packed with human purpose and commerce, just as it has since Europe and Asia began meeting here thousands of years ago.

Istanbul has so much history that it lies in heaps upon itself. It is impossible to even begin to define what this city has been and what it means, for so much has happened here over such a long time that a single human living in a single time can only begin to comprehend its endless iteration. The centre of empire through a half dozen major eras in human civilization, this city has had a profound influence not just on the region but on the world.

We turned off the crowded coastal freeway on to Atatürk Caddesi (Atatürk Avenue) and entered the Sultanahmet, one of the most famous of all the historic districts in Eurasia. While our taxi drivers tried to find our small hotel tucked away on a side street, we revelled in the dizzying atmosphere of the crowded warren of ancient streets that would soon be ours to explore.

The Hotel Accura, we quickly discovered, is so close to the Blue Mosque that the mosque's minarets could be seen along with its huge dome from the window of my second-floor room. The mosque possesses such a powerful presence that all views and perspectives in the Sultanahmet appear to align with it. Not a half hour after we checked in and met with fellow Canadian Richard Kellow, we were exploring its outer walls and the districts that fall away from the pinnacle it represented in empire, architecture and faith.

The Blue Mosque takes its name from the mainly blue Iznik ceramic tiles that decorate its interior. Iznik is one of two major regional centres where fine painted and glazed pottery was created during the Ottoman period, when Sultan Ahmet I commissioned this great monument. Some 21,043 Iznik tiles were commissioned for 50 decorative designs in the mosque, which was constructed under the direction of architect Mehmet Ağa between 1609 and 1616. Legend has it that 400 years ago the impressive scale and splendour of the imperial architect's design provoked considerable hostility because its six minarets were seen as a sacrilegious attempt to rival the architecture of Mecca itself.

The mosque is of such a scale that it is difficult to comprehend. We walked around it but were unable to find a vantage from which we could contemplate it in its entirety. We did, however, discover that impressive as it is, the Blue Mosque is not the only architectural wonder in the Sultanahmet. Nearly adjacent is the Hagia Sophia, the Church of Wisdom. Eight hundred years older than the mosque, the Hagia Sophia is the ultimate surviving testament to the sophistication, taste and religious ideals of Byzantine Constantinople. The current church, inaugurated by the Emperor Justinian in 537 CE, was built over two earlier churches, dating from 360 and 415 respectively, that burned down, implying that powerful kings have been using religion on this site to justify the ambitions of empire for more than 1,600 years.

The Hagia Sophia is so old that, over time, it qualified for complete makeovers, just as buildings in Rome did. In the 15th century, when the church site attracted the attention of the Ottomans, minarets, tombs and fountains were added. Earthquakes and time have also had their effect. The walls have been buttressed to keep them upright. Work continues today on this great heap of crumbling splendour. But the building is worth saving, for it is the kind of structure that would make any civilization proud.

Adjacent to the Blue Mosque is a large park in which stand two very large obelisks. The Egyptian Obelisk was built in 1500 BCE and stood outside Luxor until Constantine brought it to Constantinople. Next to it is the Serpentine Column, which apparently dates from 479 BCE. It was brought to Constantinople from Delphi. These obelisks stand in the ruins of a stadium that once formed the heart of Byzantine Constantinople. Like everything else from that period, the Hippodrome began as an earlier architectural form and was expanded by imperial order during the heyday of Byzantine rule. The original stadium was laid out by Emperor Septimius Severus during the rebuilding of the city in the third century CE. Constantine I expanded the size of the Hippodrome and connected it directly to his palace. During the second part of his reign, between 324 and 337 CE, the Hippodrome held up to 100,000 people, making it as large as all but the very largest stadiums that exist in the world today. We only get a hint of what it must have been like. The modern road that circles the square almost exactly follows the line of the chariot races.

After wandering around somewhat aimlessly, as one tends to do when part of a tired and jet-lagged group, we settled on a restaurant where we enjoyed our first taste of authentic Turkish food. I had kabobs and rice while the others had chicken. We saw elegant and fashionably dressed Turkish women out with their husbands and children and listened to Turkish music. Istanbul is just as cosmopolitan today as Constantinople was a thousand years ago. As Rafi Zabor wrote in a *Harper's* article that so excited me about coming to Turkey, Istanbul is in many ways the best of both worlds, a most productive overlap of Europe and the East:

I began to see ... in a series of multiple exposures or palimpsests superimposed upon the vitality of the harbor, a general representation of the million million lives that had come and gone to and from this city in wave upon wave: the Byzantines in their tiled and golden height; the Turks and their generations across the Ottoman centuries, conquest and culture, long decline, the Tulip Age, palace revolutions, teeming multitudes across these hills and waters, costume after costume, humanly faceted face after face, each pair of eyes bright with the energies of its unrepeatable life. I'd read enough history to make out some of the detail in this crowd, and it was an incredible, super-populated spectacle, all of it gilded by the majesty of the setting: Constantinopolis, Istanbul, they changed the name, but it is the one and only city of the great mixed world.

I am here, finally, 30 years later, comfortable at last in that great mixed world.

I keep the window of my room open so I can listen to the call to prayer broadcast from the great dome of the Blue Mosque. The call is so loud and clear that I think it is emanating not from the mosque, but from inside my head. It is as though I have somehow found myself next to a great river of thought that has been flowing through deeper human consciousness for thousands of years. It is as if I have found myself on the banks of some sort of Euphrates of the heart whose headwaters are the sources of our species' oldest beliefs.

The next morning, I walked to the Blue Mosque and observed again the dimensions of the Hippodrome before passing the Hagia Sophia on my way to our morning destination, the remarkable Topkapi Palace.

It is interesting to observe how long a powerful social, political and economic organizing principle can endure, particularly if it is prepared to use force to perpetuate itself. Perhaps the most persistent of these forms are the various personal and family empires that have emerged throughout the history of civilization, whether as royal families, imperial dynasties or sultanates.

The Ottoman Empire rivalled any comparable imperial concentration of wealth and power in human history. This empire came into existence with Mehmed II in 1451 and survived through some 30 sultans and 470 years until Abdülmecid II relinquished the Ottoman throne in 1924. But if Topkapi Palace is any indication, it was a grand run, with plenty of intrigue, court politics, palace revolt, concubine jealousy, rebel invasion, imperial wars and all the other trials and tribulations of trying to hold a disparate and often unwilling empire together through frequent periods of turbulent and often violent change.

The history of the magnificent Topkapi Palace begins with Mehmed II and his conquest of Constantinople in 1453. The city's strategic military and commercial importance prompted Mehmed to move the capital of his rapidly growing empire to Constantinople. In so doing he ensured that it remained the most important city in the Mediterranean region. Mehmed renamed Constantinople as Istanbul in 1453 and quickly used the city as a vehicle for consolidating sovereignty over the Balkans and the entire eastern Mediterranean. Syria and Egypt fell to the Ottoman Empire in 1516–17, bringing the holy cities of Mecca and Medina under control. By the mid 1500s, the Ottoman sultan was the central figure of the entire Sunni Muslim world. Though often associated with the excessive opulence of the kind we find at Topkapi, with its titillating and exotic stories of huge harems, the Ottoman sultans commandeered stupendous military might. They were also greatly respected for efficient administration, religious tolerance and support of poetry and the arts.

After Mehmed made Istanbul his capital, he began work on a palace worthy of his ambitions as both sultan and progenitor of future dynasties. After establishing temporary quarters on a site now occupied by the University of Istanbul, Mehmed began construction of his New Palace, which later came to be known as Topkapi, located on a promontory looking out across the Marmara to Asia. The first phase of construction was completed in 1478. The palace was built over the site of the ancient acropolis of Byzantium, upon which there remained the ruins of a number of 13th-century palaces and churches. As was common practice throughout the ancient world, the *spoglia*, or remains, of buildings that had once occupied the Byzantium acropolis provided building material for what became Topkapi.

How symbolic Topkapi's location would soon become. Imagine the Muslim world looking back from its new foothold in Europe over the blue waters of the Marmara to all of Asia. It is impossible on these grounds not to sense how culture can be a seed from which ideas might emanate in ways that would allow human society to organize itself around commerce, religion and a stable order that would make progress possible. Topkapi would become the nucleus of a new civilization. It was within its petition halls and administrative offices that the genes of empire combined and recombined into new forms and even completely new orders of ideas that the sultan would sometimes have to forcibly radiate throughout the expanding Ottoman world.

I then walked through the Bâb-ı Hümâyun, the Imperial Gate, where it was impossible not to feel immediately the serenity of the palace grounds. During the time the sultan lived at Topkapi, it was forbidden to enter armed. Weapons were displayed in the vestibule to remind visitors that the palace was a place of peace. Then, as now, there were no assurances in the world. After passing down an avenue of plane trees we arrived at the Middle Gate. The gate was guarded by soldiers with automatic weapons. But the threat was not to the sultan, but to visitors and to Topkapi itself. The Uzis the soldiers carried were a far cry from the swords and muskets that would have been used to protect the sultan from rebels and foreign invaders. But their purpose was the same.

The Turks have always had a passion for weapons. Living at the crossroads of the world, it was wise to know how to defend yourself. One of the first displays we examined exhibited weapons and armour that had been in service through the long history during which the region was controlled by the Ottomans. There were beautifully made swords, bone-bashing devices, bows and arrows and chain mail. Most of the weapons looked very heavy, suggesting that many an outcome was determined by whoever was the last man able to raise a sword to vanquish an exhausted foe. A lot of people must have taken mortal cuts in the game of empire presided over by the Ottomans. The hurt of battle undoubtedly lasted long after each war.

Passing through the Gate of Felicity, you come into the Third Courtyard, where you can peer into the magnificent Chamber of Petitions, the sultan's audience hall. Here, if you were well connected, you might be granted the closest thing to a private audience you could get with a sultan. It was in the confines of this chamber that you might, if you were forward-thinking, petition for a dam on the Euphrates or for official support for a plan to integrate the management of an entire watershed. Or at least that's the kind of thing we would have petitioned for in Canada.

Of all the legends associated with the Ottomans, there is one that still captures the public imagination: the idea that a man might have hundreds of wives and concubines. Amazing as it seems, that is exactly what the Ottoman sultans appear to have done, though there is dispute among historians as to exactly how this system was managed. We wandered through an exhibition of the sultan's Chinese porcelain collection, visited the Tulip Garden and savoured the view across the Golden Horn until at last we were able to tour the Harem and the private quarters of the sultan.

One enters the Harem from the Carriage Gate. This complex is composed of the Black Eunuch's Courtyard; the Courtyard of the Concubines, or female slaves; the queen mother's quarters; and the private rooms of the sultan himself. Evidence suggests that though the size and activity of the imperial harem changed over time depending on the interests and tastes of the sultan, the traditional organization remained essentially the same. The interesting truth about the harem was that it was far more than the pit of opulent pleasure and sexual delight that Western males might like to imagine. The imperial harem, according to historian Claire Karaz, served as a training ground for future female slaves belonging to the sultan and as a pool of potential wives for senior administrators, who often started out as pages in the sultan's court. While women in the harem obviously vied for the sexual attention of the sultan, they were also well educated and trained in a wide range of duties while they were confined to court. For the great majority of concubines, the imperial harem was a way out of the miseries of slavery.

An elaborate system not unlike today's professional sports scouting ensured that the empire was scoured regularly for harem candidates. Young girls were spotted in the slave markets, and if they were particularly beautiful or talented they were presented to the head eunuch or a court vizier, who would evaluate their suitability for harem life. The most prized women, apparently, were Greek, Italian, French, Ukrainian and Caucasian, suggesting that slavery may have once been a little more widespread in this part of the world than we normally suppose.

The training of these girls was rigorously undertaken by expert older women who had entered the harem as slaves themselves. By the mid-16th century the Queen Mother herself officiated over instruction, which included reading, writing and proper Ottoman Turkish. Embroidery was a harem specialty, which was sold through Jewish tradeswomen. Many of the girls became fine musicians, singers and seamstresses. By the age of 16 or 17 they had either caught the eye of the sultan or been married off to royal pages or other, higher-ranking imperial officials. They were provided with dowries or cash and went out into the world to become, in many cases, the spouses of wealthy and influential aristocrats in the Ottoman Empire. The practice of the harem came to an end with the "Young Turk" revolution of 1908, when Sultan Abdul Hamid II was driven from Istanbul and forced to free his black eunuchs and harem girls.

This was a great deal to take in, and I struggled to ascribe meaning and value to what I was seeing and hearing. First was the problem of allowing ourselves to be enslaved through empire. There was a painting of an imperial ceremony in the gallery exhibiting the art collected by a long succession of Ottoman sultans that showed a line of soldiers in perfect formation honouring their emperor. They lived and died by their allegiance to their ruler. The image made me wonder what makes us want to act like this as a species. What makes us want to create and submit to forces that have complete power over us? Why do we give in so utterly to structures that rob us of our freedom? What is the biological imperative for self-slavery?

I couldn't stop thinking about life in the harem and all the questions that were not being asked or answered on our tour. Sultans appear to have kept harems of varying sizes for some 450 years. They would go there to choose the women with whom they would take their pleasure. The mere visit of the sultan must have caused a stampede of jealousy and fierce competition among the women. Even the Queen Mother and a hundred black eunuchs would not be able to put down the squabbles and backstabbing that would ensue the moment the sultan had left with his choice. Out of all the pettiness and mean-spiritedness, however, the sultans achieved two things. They allowed beautiful and talented slave girls a way out of the bondage the sultans had imposed, and created a "concubine class" worthy of marrying court officials. The sultans also created the opportunity to have sex with almost an endless array of young and attractive women without the usual morning-after headaches of heartbreak and failed expectation. No wonder so many males fantasize about the harem. But while we may attempt to dignify its role and function, it must be remembered that it was in essence an act of legalizing slavery and prostitution. That it operated as a tradition and custom for nearly 500 years does not change that fact.

With the exception of the Harem, the sultan's quarters were very similar in style to the kind of opulence that existed in palaces in the capitals of other empires centred in Austria, France, England or China. The sultan was, after all, just another emperor. And that, I thought, was where this tour was likely to end. But the tour, like Turkey itself, was full of surprises.

Near the end of the route there is a grand hall occupied by two very large clocks, both with hands stopped at 9:05. The guide explained that 9:05 a.m. on November 10 was commemorated throughout Turkey as the anniversary of the death of Atatürk, the country's greatest national hero. Atatürk, we discovered later and throughout the Rosenberg Forum, was a very important figure who still managed to wield huge influence in Turkey from beyond the grave.

A proper understanding of Atatürk's contribution to history is essential to understanding modern Turkey. In 1912 and 1913, the Ottoman Empire lost most of its remaining European possessions in the Balkan Wars. In 1914 it entered the First World War on the side of Germany and Austro-Hungary. It was a costly mistake, both economically and militarily. By the time the war ended in 1918, only the heartland of Anatolia, including the cities of Antep and Urfa, which we were to soon visit with the forum, were still in Ottoman hands. Foreign troops occupied Istanbul, the coast city of Izmir and many other Turkish cities.

The occupation of Turkey by British, French and Italian troops fuelled Turkish nationalism. When Greek troops occupying Izmir began pushing toward Ankara on May 15, 1919, a bitter war ensued. The Turks were losing the war until an army officer named Mustafa Kemal took control. Kemal was a seasoned and respected veteran of the Gallipoli campaign in

1915–16, one of the bloodiest engagements of the war. After the Greeks were routed at the battle of Dumlupinar and all foreign forces expelled from Turkey, a new republic was proclaimed. The Treaty of Lausanne in 1923 recognized the country's new borders and territories, and Ankara became the state's new capital. Mustafa Kemal was, not surprisingly, elected the leader of the revived state. An admirer of European institutions, Kemal adopted legal codes used in Germany, Italy and Switzerland and dissolved the caliphate, the traditional ruling body. In 1928 Turkey was declared a secular state with a European-style constitution. Islamic courts and religious schools were abolished and Arabic and Persian alphabets were replaced by the Latin-based alphabet of the West. After 1935 Mustafa Kemal became widely known as the "father of the state," or Atatürk. When he died in 1938, the foundation for Turkey as we know it today had been established. There is good reason why, at 9:05 a.m. on November 10 of each year, Turks all over the country stop what they are doing, whether talking on the phone or driving their car, to offer a minute or two of silence in honour of Atatürk, the father of modern Turkey.

After returning briefly to our hotel, I joined my colleagues for lunch at an outdoor café on a crowded street near the Blue Mosque. While we were talking and eating, a man from an adjacent table came over and explained he had just arrived from Baghdad recently with his new Iraqi wife and had heard us speaking English. At first I thought this man may have simply been an articulate carpet salesman. As it turned out, however, he was an American engineer who had gone to Iraq to help rebuild the country's transportation system. He introduced us to a thin, hollow-eyed woman who was his new wife. She had the look of someone who had seen more than she could ever bear or process either intellectually or emotionally, a look that spoke of unimaginable pain. The man, who looked like he might be in his 50s, explained it was unusual for them not to be hearing gunfire. We did what we could to wish the woman and her new husband luck, but I am not sure she was listening. Like Iraq itself, she seemed on the verge of imploding, of collapsing inward toward ineluctable contradiction. Like the others in our party, I ached for her. Amidst all this human heartbreak one can only wonder what is happening to the water resources in that afflicted country.

2. GAZIANTEP

Water has been critical to the making of human history. It has shaped institutions, destroyed cities, set limits to expansion, brought feast and famine, carried goods to market, washed away sickness, divided nations, inspired the worship and beseeching of gods, given philosophers a metaphor for existence, and disposed of garbage. To write history without putting any water in it is to leave out a large part of the story. Human experience has not been so dry as that.

— Donald Worster, *Rivers of Empire*

Because the prestigious invitation to participate in the Rosenberg International Forum on Water Policy was predicated on our availability for the two-day preconference field trip through southeastern Anatolia, we left for the domestic airport hours in advance to ensure we wouldn't miss this important connecting flight. As we lingered in the departure lounge, participants from all over the world began to trickle in.

It was getting dark as we approached Gaziantep, but even in

the gathering gloom we could see how lushly green it was, especially compared to the arid regions we had flown over for an hour as we traversed most of Turkey from north to south. All this lushness was a gift from the massive Southeastern Anatolia Project, known locally as GAP. It was this integrated hydro power, irrigation and economic development initiative that we were in southern Turkey to see and discuss.

The inimitable Rosenberg co-chair Dr. Henry Vaux surprised everyone by being on the tarmac as the plane nosed to a stop in front of the terminal. Around him were plainclothes and uniformed police whose role was to ensure we all got our luggage, bypassed passport control and boarded a bus which would be escorted to the hotel in Şanlıurfa by a pair of police cars with their lights flashing. This would be the kind of security we would enjoy for the rest of our visit to southeastern Anatolia.

Over dinner at the Tuğcan Hotel, I began to appreciate the depth and experience of the Rosenberg participants. One of them was a very sophisticated and articulate Thai. When I asked him what he did, he explained he was a student of the Mekong River. When pressed further, he told me he had been pursuing this expertise all his working life. He then lamented the absence of a functional civil society that would care for his beloved river. Dams were ruining the Mekong. He was concerned that freshwater dolphins and the amazing 300-kilogram Mekong catfish were virtually extinct and that the dams would push them over the brink.

At the same table was an expert on that part of the Colorado River that flows diminished and polluted into Mexico. He explained that the low flow actually reaching the ocean has caused a reversed estuary in terms of dilution. Sea water was flowing farther and farther upstream into the river. In association with this, however, freshwater organisms were demonstrating surprising adaptive capacity in responding to the increasing salinity.

Also at the round table was an Indian working in Sri Lanka. He asserted we needed to use knowledge better to enable us to move on from positional arguments and get to common interest so that we can come to decisions in time to make a difference in water quality and availability issues. He also said trust played a crucial role in moving people from positions to common interests. There has to be someone that everyone trusts in water management, he maintained, if advancement is to be possible.

I also happened to listen in to a very stimulating conversation between an Australian, an Englishman working with the World Wildlife Fund in Hungary, and a senior government official responsible for water management. The Australian argued that the greatest intellectual challenge of our time was to make the leap from environmentalism to true sustainability. "We need to make decisions that optimize social, cultural and economic sustainability in the context of an enduring environment," he said. "In other words, we can no longer afford to make decisions that reflect only short-term political and economic considerations."

This was only the first evening. Given the intensity of the conversation, I couldn't get over my good fortune in being invited to attend this elite forum. I could see already that this was the type of experience that could change one's life. As the discussion continued I happened to glance out the window and saw a trio of sheep being driven down the sidewalk in front of the hotel. Worlds really do come together in Turkey.

I concluded from these preliminary talks that the Rosenberg Forum is a metaphor for what we need to do. We need to gather together experts from the world over and hear one another. We must change the institutions upon which we rely to define our future. We must do so now.

After breakfast the next morning, we gathered for orientation in one of the hotel's meeting rooms. Henry Vaux introduced our group to the head of the regional and transboundary waters department in the Turkish foreign ministry, who was to offer an introduction to the geography, geology, hydrology and history of southeastern Turkey. In his presentation, he offered that there were many transboundary water issues, as Turkey was both an upstream and a downstream riparian state. Water flows into Turkey from three countries: Bulgaria, Lebanon and Georgia. Both the Tigris and Euphrates have their origins in Turkey, which has implications for Syria and Iraq, as they are downstream nations. Because of the complex nature of these issues, the management of water very much fell under the Ministry of Foreign Affairs.

A concise historical summary followed which told much of the grand story of the classical ancient world. The area around what is now Gaziantep was settled by the Hittites around 2800 BCE. The Hittites were in turn conquered by the Persians. In 333 BCE Alexander the Great appeared on the Aegean and defeated Darius and the Persians before proceeding south to take Cairo. The Assyrians were in power next, then the Seljuk Turks and then the Ottomans. There was a brief attack by the French in 1921, which was repelled by Mustafa Kemal, who, as mentioned earlier, later assumed the name Atatürk, the father of the modern Turkish state.

It was then offered that Gaziantep was the centre of a

multi-sector, integrated river basin management project which the Turks were very proud of. It was mentioned that what was being done in Gaziantep was a model of sustainability. A total of US$16-billion had already been spent on the Southeastern Anatolia Project, an undertaking which Turkey hoped would ultimately restore this cradle of human civilization to its former glory. We would spend the next two days touring the project to determine whether GAP was indeed a model of sustainability and these high hopes were suitably placed.

Henry Vaux then rose to explain the objectives of the pre-conference field trip in the context of the larger Rosenberg Forum goals. The first object of the trip was for participants to get to know one another. The second was to gain a full and detailed understanding of what was working and what was not in terms of transboundary water management. The third goal was to enable participants to see some of the antiquities of upper Mesopotamia. All right, I said to myself, if there are antiquities, I'm in.

Our first stop on the field trip was at the Gaziantep Archaeological Museum, to see mosaics rescued from the ancient city of Zeugma before dam construction was begun on the Euphrates and the site flooded. The police stopped all traffic so that our bus could park right in front of the entrance gate. A dozen police and security officials cleared the way for our exclusive entrance. As one of the participants from Spain pointed out, it was hard not to feel safe.

Zeugma was one of four important cities in the Kingdom of Kommagene. First known as "Seleukeia of Euphrates" during the Hellenistic era, this old city was a commercial centre on the earliest Silk Route from Antioch to China. Located on the banks of the Euphrates, this centre grew prosperous enough to afford its own artistic traditions, which included architecture and mosaics. During the Roman era a military unit consisting of Anatolian soldiers was stationed there. This unit, called the Scythian Legion, gradually acquired an increasingly Roman character and began to effect a Romanization of the region. The Roman influence began to manifest itself in statuary, steles (inscribed or sculpted upright slabs or pillars), rock reliefs and formal altars. The town later became known as Zeugma and was soon connected by a bridge, for which it is named, to the city of Apameia on the other side of the Euphrates.

Though it remained a modest settlement through the 11th and 12th centuries, Zeugma disappeared from the records, hinting that it had been completely abandoned to wind and time. It was not until 500 years later that the village of Belkis was established in the same area. The first excavation of Zeugma was undertaken in 1987 by a team from Gaziantep Museum who unearthed graves in the necropolis of the ancient city.

When it became clear that the proposed damming of the Euphrates threatened sites of ancient communities, an archaeological program was initiated to inventory areas along the river projected to be flooded. In 1992 the Turkish ministry of culture began excavating the site of Zeugma using side-scan radar. The archaeology advanced quickly, as the site was scheduled to be submerged under the reservoir that was soon to be created behind the Birecik Dam. Zeugma suddenly became more than just a bridge over the Euphrates. It was a bridge from the past to the present, at least in terms of art.

From June to October of 2000, Zeugma was the site of one of the most ambitious excavations and archaeological rescues in history. More than a hundred specialists from Turkey, Great Britain, France and Italy employed some 250 workers at 19 different sites. During excavations conducted by the Gaziantep Museum, a floor mosaic depicting the wedding of Dionysus was found in an almost perfect state of preservation in an unearthed villa. Unfortunately, many of the frescoes and mosaics archaeologists discovered could not be moved. These were covered with a special plaster which will supposedly protect them from the effects of the water. Anything that could be moved was carefully dug out and taken to the Zeugma excavation center. Transportable items included decorative iron window cases, figurines, glass objects, bronze statues, coins, helmets, lances, knives, gold rings, gold leaf, column bases and steles depicting Antiochus, the king of Kommagene, shaking hands with Apollo. There were also three very significant mosaics: *The Kidnapping of Europa*, *Eros and Psyche*, and *Three Women*, which are presently being restored.

There was also one other mosaic fragment, a depiction of a young girl, that in my opinion is a real masterpiece. I have come to call this work the Anatolian *Mona Lisa*. What is amazing about it is its haunting depth effect. While the other mosaics depict interesting scenes, they did not leap out in the same marvellous way as this incomplete but wholly engaging portrait of a young woman whose eyes, though composed of nothing more than bits of coloured stone, not only appear to follow you around the room, but do so three-dimensionally.

In Florence, visitors to the Uffizi Gallery are told that pictorial art had been two-dimensional in representation until Cimabue began to experiment with the kind of shadowing that could make flat surfaces appear to stand out in relief. Then Giotto burst through that door cracked open by Cimabue to create the techniques artists still use today to give depth to their pictorial works. But as significant as this development in art

history was, the Anatolian *Mona Lisa* demonstrates that it already had quite an ancestor. The haunting eyes of this portrait show us that 700 years earlier, some 30 generations before this same effect was discovered by Cimabue, an artist in a backwater military town on the Euphrates in what is now Turkey had figured out how to arrange small chunks of coloured rock to create the illusion of depth.

I have often wondered if the Garden of Eden was a real place, and if it was, what happened to it. One of our objectives over the two days of this field trip was to try to understand the progression of human experience in the cradle of Western civilization, the place in the world where people have lived in settlements the longest. Here in upper Mesopotamia, we can observe layer upon layer of human presence and impact. We are not on a tourist junket, for we are accompanied by experts in the history of the region, hydrologists, engineers and economic planners who want to know what the participants in this forum think about what they are trying to do in southeastern Anatolia. They want us to see what the area might have been like as evidenced by its antiquities, and what upper Mesopotamia is like now as expressed by the modern Turkish state. They want us to help them imagine how the region might restore itself to its original state through the massive GAP project, which seeks to employ the Euphrates to irrigate a tired and desiccated landscape back to its original productivity. Though no one has mentioned it, what we are seeing is profound. We are exploring what we humans did to the Garden of Eden, and what is possible to do today to restore that condition to Mesopotamia and the world.

It would be understandable if our generous Turkish hosts preferred that our focus fall principally on what they have so proudly accomplished: the economic and social redevelopment of an entire region by means of huge hydro power and irrigation projects created under the GAP program. This diverse group of global water experts, however, will likely be somewhat hard to control. They tend to let their interests gravitate toward comparative analysis. They have seen a great deal, and they want to use this extraordinary opportunity to put that experience into even more productive relief. They are not the type of people who will take public relations language and spin tactics at face value. They judge projects of this kind not by what proponents say but by what they see with their own eyes. There is an undercurrent within this group that suggests they have seen all this before. Though they are impeccably polite, I get the sense that many of them are unmoved by the constant repetition of the economic statistics related to the project. I get the feeling in talking with them that many of them believe huge dam projects like these are not going to bring anything but temporary prosperity to southeastern Anatolia. Maybe I am reading them wrong, but they don't seem convinced there is much about this approach that hasn't been tried before. There is a sense that what brought the demise of the world's first agrarian culture, on the banks of this very river, will bring down the present culture too, only it will not take nearly so long to do it.

The Turks have a different view of their project. They believe in the economic miracle they are creating here. They want us to look beyond any preconceptions we might have about the dams, in order to see instead the actual jobs and prosperity they are focused on. They believe that what they are doing is sustainable. After visiting the museum and its artifacts rescued from the advancing waters of the Birecik reservoir, we were convoyed by our police escort to the Gaziantep Organized Industrial Zone, where a great deal was done to illustrate the way economic development was planned into the GAP project right from its inception.

In order to understand the full implications of how careful use of water resources was being employed as an economic engine for revival of prosperity in this region, we visited two large factories, one making plastic bags, mostly for the European retail market, the other turning out cotton textiles.

According to materials we received in advance of the forum and to the presentations we attended before visiting the Gaziantep industrial park, the Southeastern Anatolia Project was initiated in 1976 as a large-scale, multi-sectoral regional development that aimed to bring this region of Turkey into the modern economic world. Some 13 major multi-sectoral projects fall under the GAP umbrella.

The scale of the GAP project is hard to grasp. It involves the construction of 22 dams and 19 hydroelectric plants. The object of this massive development is to generate cheap power that will drive manufacturing investment such as the plastics and textile plants we visited in the greater and quickly expanding Gaziantep Organized Industrial Zone. This appears to be doing exactly what was intended. In Gaziantep at least, employment rose from 20,000 jobs to more than 150,000 since GAP began, based almost solely on manufacturing and services attracted to the area by the abundant availability of inexpensive electricity.

It was clear we were only seeing a very small part of what was happening in terms of manufacturing in the Gaziantep industrial zone. This zone, we were told, currently produces 70 per cent of the machine-made carpets in all of Turkey. It produces 60 per cent of its acrylic yarn, 60 per cent of the country's macaroni, 25 per cent of its wheat flour, 60 per cent of its polypropylene and 20 per cent of Turkey's PVC construction materials. It also exports many of these materials to some 119 countries, which would account to some extent for the heavy truck traffic on highway E90.

What is manufactured in this zone represents 5 per cent of the entire industrial production of Turkey, and all of it is made possible by cheap power from GAP hydro projects. The great success of projects like this, however, is creating problems of its own. Average energy consumption in the Gaziantep Organized Industrial Zone is rising at about 5.7 per cent a year. But because of population growth, urbanization and industrial development, total energy consumption in Turkey as a nation is rising at 20 per cent a year, more than three times the rate of increase in Gaziantep. The GAP project appears to have anticipated this growth. According to the plan, this is just the start of a bright future. At full build-out the GAP project is expected to generate a staggering 27.3 billion kilowatt hours of electricity annually, which if harnessed productively could make southeastern Anatolia the modern economic engine of a revitalized Turkey. But, as astounding as it currently is, the plan does not stop there.

The GAP project also plans to provide irrigation for 1.7 million hectares of farmland, or about one-fifth of all the irrigable land in Turkey. The project also aims for total integration of all relevant sectors of Turkey's economy and culture, including transportation, social infrastructure and cultural development through targeted regional educational programming. As befits a project of social and economic engineering on this scale, GAP is a top priority of the Turkish government, administered by the prime minister's office itself. Turkey is counting on this project to bring the country fully into the Western world and the global economy. Through large-scale innovation of this calibre, Turkey hopes to qualify for entry into the European Union and grow to become the prosperous and powerful state its leaders and people feel it deserves to be. And all of this will be done through the careful management of one single, simple resource: water.

So what could possibly be wrong with a large-scale megaproject, supported by the state, that aims to bring prosperity and hope to the tired lands of upper Mesopotamia? What could possibly go wrong with a project into which so much intelligence and such huge financial resources have been invested? How could we, as outsiders, find anything critical to say about what the Turks are trying to do here? The purpose of this field trip was to come to terms with questions such as these and to pose others.

Only a day into the trip I was beginning to sense the utter enormity of the considerations the Southeastern Anatolia Project posed. The questions we had before us at the forum were as big as the GAP project itself. Bigger even, for they confronted us with the broader question of how to use water in a sustainable way in the service of humanity, not just in Turkey but everywhere in the world. These questions also underscored a deeper accountability: the responsibility of experts in any field to share their experience, perspectives and knowledge with one another even if it is painful to do so. And though I didn't have a fraction of the experience of others in this forum, I was nevertheless beginning to see, with each kilometre we travelled, that the Turks seemed to be making the same kinds of mistakes here that have been made elsewhere, including in places where forum participants lived and worked.

After we'd finished a very generous lunch at the main office of the industrial zone, our police escort cleared the way for our motorcoach to proceed east on E90, the crowded southern highway that crosses Turkey from the Mediterranean. An endless line of trucks on this route hauled freight to Syria and war-torn Iraq.

We had our first stunning look at the Euphrates at Birecik, where we stopped to visit a recovery and breeding centre the GAP project had created for the bald ibis. This species had begun to suffer in the 1970s as a result of agricultural use of pesticides, which wiped out the insects that had formed the diet of this remarkable bird. As the poisons destroyed its food, the ibis was starved into near extinction. What was important about the recovery centre, our hosts explained, was that the prosperity brought about by major integrated development of hydro power resources made it possible to fund important environmental programs that benefited not just Turkey but all of the countries in the range of a species like the bald ibis. The point about it costing money to preserve endangered species was unassailable. It does take prosperity to be able to do this. The hydro power and economic development authority for this region was clearly trying hard to be all things to all people and break through our contemporary organizational and ideological barriers into a truly sustainable way of thinking about and acting on the future. This is not a simple act, however. While the Turks must be congratulated on providing for the ibis, what we were seeing in southeastern Anatolia did not strike me as being any more sustainable than what I see at home in Alberta. Still, what was happening at the ibis recovery center was important in demonstrating that large-scale, multi-sectoral economic development schemes like GAP could also embrace conservation, and that is the first step down the road to sustainability.

I must admit, though, I had trouble concentrating even on these exotic birds with the river beckoning at my back. It was the fabled Euphrates that I had come to Turkey to see, and there it was, only a few metres away. I was immediately impressed with its size. While everyone on the bus would have been comparing it with the rivers they knew and loved most, I could not measure it against any prairie river I was familiar with. It was too wide and imposing relative to anything that flowed down the eastern slopes of the Canadian Rockies. The only river I could compare it to in western Canada was the Columbia at the spot where the Trans-Canada Highway crosses it at Revelstoke.

While the others continued to concentrate on the ibis and its dietary problems, I walked across the road and down to the river, where I took off my shoes and waded right in. The cool water seemed to connect me with the nearly 500 generations of people who have lived along these banks. And who among all those millions would not have refreshed themselves on a hot afternoon in exactly this way? I had prepared for months for this moment. I think of water and I think of thirst. What else is there to think of in this dry land?

Standing in the waters of this most famous of rivers, I was reminded of Donald Worster's opening remarks in his landmark book *Rivers of Empire*:

> The fear of going dry has driven many communities to extraordinary efforts, provoking in them the deepest anxiety, the sorriest desperation, forcing them to make radical changes in their behavior and institutions. It has stirred them out of lethargy to undertake the most difficult labors: building enormous engineering works to bring water from distant places and stave off their thirst. That reaching out to establish control over a river, often driven by a raw instinct to survive, has had profound implications for the course of history. In light of such human endeavour, history has become no longer a matter of Euphrates dominating people, but of people bent on dominating Euphrates.

Worster explains that three broad modes of water control have appeared so far in history. Each had its own complex of techniques and apparatus, its own pattern of social relationships and arrangement of power. The first of these was local subsistence agriculture, which employed primitive forms of irrigation on a small-group scale. The second was the agrarian state model as practised on the banks of this river for thousands of years. And finally there is the capitalist state model as used today in the American West and now here on the Euphrates.

We all know what subsistence irrigation looks like, for it is still practised widely in the developing world. We are also familiar with the hard-edged, bottom-line-oriented industrial-style irrigation that is presently being employed throughout much of western North America. As Mesopotamia has changed almost beyond recognition since the world's first agrarian society developed here some 11,000 years ago, we have to rely on historians like Worster to describe the basic elements of such a society. In his description it was this form of agriculture that made wholesale concentration of power possible:

> The state provided an adequate and dependable supply of water to the village, and in turn demanded a payment of tribute in the form of money or crops. A new redistributive economy thus appeared, wherein wealth flowed from the outlying village to the capital city and then, as expenditures for water engineering and maintenance, back outward again. Always, however, a large part of the wealth stayed in the capital city, where it paid for luxurious homes for a new ruling class or for standing armies to defend the irrigation society against its enemies, usually marauding nomads. Given enough tribute, which conversely meant given enough water supplied to the villages, the rulers could create an empire. And that is precisely what many of them did. Each time they extended their canals into new territory, they added to their domain, and, in turn, increased their tribute until at last their domain extended well beyond any conceivable gift of water. In those desert empires, the shape of power, therefore, was like that of some primitive marine animal: a vast amorphous tissue of villages, weak and disorganized, dominated by a more highly evolved central nervous

system. [Karl] Wittfogel called this animal a hydraulic society. But to make matters clearer, since I will argue that hydraulic societies come in more than his one variety, we can call this second type the agrarian state.

It is widely held that the first true agrarian state developed here on the banks of the Euphrates. The Fertile Crescent, which extended through southeastern Anatolia following the course of the Tigris–Euphrates river system through what is now Syria and Iraq, was perhaps the earliest centre of food production in the world. It is also the origin of several of the modern world's major crops and almost all of its domestic animals. The agrarian state was also the birthplace of a whole string of other developments, including cities and writing, which, as Donald Worster recounts, led to empires and the first true civilization.

As Jared Diamond points out in *Guns, Germs, and Steel*, all of these important developments sprang from the dense human populations, stored food surpluses, and feeding of non-farmers that were made possible by the rise of crop cultivation and animal husbandry. According to Diamond, food production was the first of the major innovations to appear in the Fertile Crescent. He commits a substantial part of his book to explaining why this region's domestication of plants and animals gave it such a head start in the development of human civilization.

The first advantage the Fertile Crescent had is the fact that it lies within the so-called Mediterranean climatic zone, with its mild, wet winters and long, hot, dry summers of just the kind we were experiencing on our tour in southeastern Anatolia. This climate, Diamond explains, selects over time for plant species able to survive the long, dry summer and the 35–40°C temperatures we were experiencing. One significant adaptation plants make to these Mediterranean conditions is that they become annuals instead of perennials. In the Mediterranean context, this means they dessicate and die during the hot dry season. With only one year of life, annual plants inevitably remain relatively small. Instead of investing their nutrient energy into growth, it they invest it in the production of big seeds. Instead of woody and fibrous stems such as those which in most places form bushes and trees, these annuals produce huge numbers of seeds. As it happens, these seeds, particularly those of cereals and a number of varieties of leguminous plants classified as pulses, are edible and cultivatable by humans. Three of these annual cereals – emmer wheat, einkorn wheat and barley – and three of the pulses – peas, lentils and chickpeas – still constitute today six of the world's dozen major crops. The wheat that has made the Canadian prairies what they are today began as an abundant annual flourishing in this very river valley.

Standing knee-deep in the Euphrates, all manner of questions come to mind. Why didn't civilization first emerge instead in the boreal forests of modern-day Canada or Russia? Diamond meditated on this same question. He concluded that while some tree species in wet climates do produce big, edible seeds, those seeds are not adapted, as grasses are, to survive long, hot summers. These adaptations are crucial to the story of civilization in that they produce precisely those dry-hardy qualities that make it possible for these grains to be stored for long periods without spoiling. Not unlike squirrels, people learned to do exactly this on the banks of the Tigris and Euphrates rivers. After learning to store seeds, we then began to grow them.

The second huge advantage of the Fertile Crescent flora, according to Diamond, was that the wild ancestors of many of the area's proto-crops were abundant and highly productive. They likely existed in large stands whose value must have been obvious and attractive to hunter–gatherers. Recent studies suggest that natural stands of wild cereals in the Fertile Crescent may have yielded up to a ton of seeds per hectare. This suggests that natural plant communities in this region may very well have yielded 50 calories of food energy for only one calorie of work expended in gathering them. No wonder this part of the world was considered a paradise. Put into today's crassly commercial terms, the food-finding return on investment in the Fertile Crescent was astounding. As supplies could be made to last naturally, and be encouraged to reoccur reliably through agricultural cultivation, you could essentially make $50 for every $1 of investment in time and energy. By collecting huge quantities of wild cereals in a short time when the seeds were ripe, and storing them for use as food throughout the year, some clever Fertile Crescent gamblers were able to kick the hunter–gatherer habit and settle down into permanent communities where they were able to imagine making the step from simply collecting abundant wild plants to encouraging their growth in a manner that would allow them to generate even greater returns on their investment of personal and collective energy and effort. In a very real sense it can be said that the economic animal we are today was born on the banks of this river. That economic animal cut his teeth on the early cultivation of wild cereals. He then spent his adolescent years establishing the first human empires on the foundation of a hydrological elite whose power was based on knowledge of how to control and profit from the river. This animal achieved some kind of imperfect maturity through the undertaking of hydraulic megaprojects of the kind we were seeing in southeastern Anatolia today.

A third advantage natural plant communities conferred upon

the Fertile Crescent, Diamond tells us, relates to the fact that these communities included a disproportionately high number of hermaphroditic, or self-pollinating, plants that on occasion also allowed themselves to be cross-pollinated. As early farmers in the Crescent had plenty of time to observe what was happening to the plants to which they had decided to hitch the wagon of their history, it didn't take long in the larger scheme of things for them to recognize that some clumps of wild cereals had qualities of endurance or seed production not possessed by others. While they didn't understand exactly why this was so, it gave them an opportunity over time to ponder simple methods of selection that might improve the characteristics of the wild plants they had begun cultivating. Cross-pollination created new varieties to choose from.

They soon observed that occasional cross-pollination occurred not only among cereals of the same species but also between related species to produce interspecific hybrids. This, Diamond explains, is a lot more important than it first appears. One of the most significant hybrids to emerge among cross-fertilizing hermaphroditic grass species was bread wheat. This humble child of the Euphrates Valley is now the most valuable crop in the modern world. The reason for the remarkable spread of this crop was its high protein content. Einkorn wheat and emmer wheat reliably possessed 8 per cent to 14 per cent protein by weight. In comparison, the most important cereal crops of Eastern Asia and the New World – rice and corn – had a much lower protein content, which under some circumstances posed significant nutritional problems for the early civilizations that came into existence by cultivating them.

Diamond goes on to ask why this particular Mediterranean climate generated the earliest civilizations instead of similar climates that exist in California, Chile, southern Africa and southwestern Australia. In considering the answer to this question, Diamond introduces the reader to some very interesting dimensions of island biogeography. The Mediterranean zone of western Eurasia enjoyed at least five significant advantages over other Mediterranean zones elsewhere in the world.

The first is that it is the largest zone of this climate on the planet. The first rule of biogeography is that the bigger an eco-region is, the greater its diversity of wild plant and animal species is likely to be.

The second advantage the Eurasian Mediterranean zone possessed was a great climatic variation from season to season and year to year. This dramatic range of variation favoured the evolution of an especially high proportion of annual plants. According to Diamond, some 32 of the world's 56 most prized grasses were found naturally and abundantly in the Fertile Crescent. These, he points out, are the cream of nature's crop-grass species, with seeds at least ten times heavier than the median grass species. In terms of protein per volume, these grasses are the mothers of all angiosperms. These are the plants Loren Eiseley visualized in some muddy hand at the beginning of the world, the great evolutionary leap in our green reality that, once grasped, transformed us from nocturnal prowlers into a civilization capable of ruling, and ruining, the planet.

The third advantage of the expansive Mediterranean zone in the Fertile Crescent is that it is topographically diverse as well. Its range of elevation is truly impressive. Within this zone is the lowest place on Earth, the Dead Sea. But there are also mountain ranges, like those near Tehran, which rise to 18,000 feet, or more than 5000 metres. Perhaps most importantly, these diverse topographies exist close to each other, creating a corresponding variety of environments and hence a broad diversity of wild plants able to serve as potential ancestors of crops. This range in altitudes also meant that hunter–gatherers could move up a mountainside harvesting grain seeds as they matured. They didn't have to play the dangerous game of having to rely on a concentrated harvest season at a single altitude where everything matured simultaneously, or maybe didn't mature at all. Farmers could also take the seeds of wild cereals growing on hillsides dependent on unpredictable rains and plant them in the damp valley bottoms where they could be watered and tended.

The fourth advantage afforded by this particular Mediterranean zone over all the others is directly related once again to biodiversity. The combination of the widest diversity of plant species, combined with the highest percentage of annuals, made it inevitable that the region should also possess the broadest diversity of animals. According to Diamond, animals suitable for domestication were few or non-existent in the Mediterranean zones of California, Chile, southwestern Australia or southern Africa. In comparison, the Fertile Crescent possessed no fewer than four species of large mammals that were domesticated early in the continuum of settlement. These included the goat, sheep, pig and cow, which were domesticated in the Euphrates region earlier than any other animal except the dog. As Diamond points out, these four animals, along with the chicken, which was domesticated later, now form the essential basis of livestock agriculture over most of the world. It is, in fact, impossible to imagine our reality without them. Clearly, there is a lot we owe the people who created the first human civilization along the banks of this river.

The final advantage this region possesses over others similar to it was that it was not a superior environment for hunter–gatherers. It did not have abundant coastal or river fisheries or large herds of seasonally migrating mammals that could be hunted reliably. Because of this, the option of settlement and food production was easier to consider. There was, as a result, less tension between these modes of being than elsewhere. Hunter–gatherers were in fact predisposed to agriculture and herding, by the very nature of Mesopotamian ecosystems. One could say that to some extent the natural environment facilitated human settlement and civilization.

Standing in the Euphrates I feel all of human history flow past me. Human agriculture was born in the Fertile Crescent by early domestication of eight "founder crops" and four species of livestock progenitors. These founder crops are so named because they were the basis of an agriculture that began here and radiated outward to the rest of the world. Of these eight species, only two – flax and barley – range in the wild at all widely outside the Fertile Crescent.

Two of these founder crops had very small natural ranges. The chickpea was confined to what is now southeastern Turkey and the very river valley in which I am presently standing, watching the Euphrates flow cool and clear over the pebbles beneath my feet. Emmer wheat was confined to a slightly larger area than what is now Turkey but did not grow outside the Fertile Crescent. In other words, these crucially important human food sources could only have been cultivated and domesticated here, because they occur in the wild nowhere else in the world.

While the "founder species" of modern livestock were domesticated in different parts of the Fertile Crescent, all four lived in sufficiently close proximity that they were readily transferred from one part to others, so that in time all four were found throughout the region. Thanks to the availability of suitable wild plants and animals, the Fertile Crescent quickly developed a diverse and balanced biological assemblage that set the stage for intensive food production worldwide. Over time, agricultural production became the foundation for a settled way of life. The industrious and forward-thinking farmers here found a way to harness the wild world and make it into an engine that reliably produced carbohydrates, protein, fat, clothing, traction and transport on demand.

A millennium later we take for granted that all these things are instantly available. We don't even think about what it took to harness the great web of life on Earth so that we could align its energies and hierarchies in such a way that they serve us and not themselves. On the foundation of agriculture we were able to build all the rest, tapping into progressively more energetic natural processes until now there doesn't seem to be a natural system left that we have not somehow begun to harness and domesticate. The ideas that first emerged here about changing the human relationship to nature radiated outward from the banks of this river to change the world. In a very real way, these ideas became the world, at least as we today think we know it.

The spread of agricultural knowledge and technique from this area to the rest of the world, it must be pointed out, was also due to a number of other fortunate natural alignments. Unlike North America, which has as its axis a north–south alignment encompassing vast differences in latitude and climate, Eurasia has an east–west axis that extends along a temperate to Mediterranean alignment with only a small range in latitude. This, according to Diamond, positioned the Fertile Crescent to launch agriculture over a broad band of temperature latitudes from Ireland to the Indus Valley. It is as if there was only one moment in time when the continents were aligned just right, when all the right plants existed in appropriate relative abundance and when the wild animals were more amenable to being domesticated by humans than they had been in past eras. It was a time when the sun shone just right and the water in the river sparkled just so. And in that perfect once-in-four-million-years moment, humans stumbled into the Euphrates Valley just in time to see the world suddenly turned into a blazing crystal whose clear light illuminated what the planet was going to be like for the next 10,000 years.

It was that same ancient inner light that guided the first settlers to the Canadian West, with their wheat and their cows and pigs. Like the light of a distant star that keeps on going long after the star that generated it has died, so the wheat and barley, the cows and pigs and the silo and plow continue on. In remote Canada, half a world away, the light of the Fertile Crescent illuminates the fields and the rangeland even though the source of that light in Mesopotamia is in ruins. I stand in the waters of the Euphrates, a traveller from an outpost of wheat and cattle created in the New World using the same plants and animals that were domesticated here. The Euphrates has flowed through me always, and now I am home. But home has changed.

So, what has happened to the Euphrates since the light of humanity's bright morning first fell on the golden stalks of emmer wheat, illuminating the future of the world? Looking up from the banks of the Euphrates, here in the cradle of the human world, I see a land that little resembles what supposedly existed here when humans first settled and began farming. Southeastern Anatolia is largely desert. The only green, the only trees, the only cool in the entire distance from Gaziantep to Birecik is along the banks of the river or the irrigation canals.

There are no clumps of wild grasses, no paradise groves of trees, no beautifully and lushly irrigated farmlands. Except along the river itself, this part of Turkey looks as though the world stopped turning somehow and this was the place that was left to bake motionless in the sun. What happened here that the land has been so altered, and what can we learn from it?

Jared Diamond pursues a very interesting line in answering this question in *Guns, Germs, and Steel.* Why is it, he asks, that in Eurasia it was the European societies rather than those of the Fertile Crescent or even China and India that colonized North and South America and Australia, took the lead in technology and became politically and economically dominant in the modern world? The question might be posed in a slightly different way in the context of our participation in the Rosenberg International Forum on Water Policy. Why is it that Turkey is doing everything it can to gain entry into the European Union rather than Europe doing everything it can to gain entry into an economic union that emerged along with one of humanity's earliest civilizations out of the Fertile Crescent? In examining these related questions, it is worthwhile to quote Diamond at some length:

> For the Fertile Crescent, the answer is clear. Once it had lost the head start that it had enjoyed thanks to its locally available concentration of domesticatable wild plants and animals, the Fertile Crescent possessed no further compelling geographic advantages. The disappearance of that head start can be traced in detail, as the western shift in powerful empires. After the rise of the Fertile Crescent states in the fourth millennium BC, the center of power initially remained in the Fertile Crescent, rotating between empires such

as those of Babylon, the Hittites, Assyria, and Persia. With the Greek conquest of all advanced societies from Greece east to India under Alexander the Great in the late fourth century BC, power finally made its first shift irrevocably westward. It shifted farther west with Rome's conquest of Greece in the second century, and after the fall of the Roman Empire it eventually moved again to western and northern Europe.

The major factor behind these shifts becomes obvious as soon as one compares the modern Fertile Crescent with ancient descriptions of it. Today, the expressions "Fertile Crescent" and "World leader in food production" are absurd. Large areas of the former Fertile Crescent are now desert, semi-desert, steppe, or heavily eroded or salinized terrain unsuited for agriculture. Today's ephemeral wealth of some of the region's nations, based on the single nonrenewable resource of oil, conceals the region's long-standing fundamental poverty and difficulty in feeding itself.

In ancient times, however, much of the Fertile Crescent and eastern Mediterranean region, including Greece, was covered with forest. The region's transformation from fertile woodland to eroded scrub or desert has been elucidated by paleobotanists and archaeologists. Its woodlands were cleared for agriculture or cut to obtain construction materials, or burned as firewood or for manufacturing plaster. Because of the low rainfall and hence low primary productivity (proportional to rainfall), regrowth of vegetation could not keep up with its destruction, especially in the presence of overgrazing by abundant goats. With tree and grass cover removed, erosion proceeded and valleys silted up, while irrigation agriculture in the low rainfall environment led to salt accumulation. These processes, which began in the Neolithic era, continued into modern time. For instance, the last forests near the ancient Nabataean capital of Petra, in modern Jordan, were felled by the Ottoman Turks during the construction of the Hejaz railroad just before World War I.

Thus, Fertile Crescent and eastern Mediterranean societies had the misfortune to arise in an ecologically fragile environment. They committed ecological suicide by destroying their own resource base. Power shifted westward as each Mediterranean society in turn undermined itself, beginning with the oldest societies, those in the east (the Fertile Crescent). Northern and western Europe has been spared this fate, not because its inhabitants have been wiser but because they have had the good luck to live in a more robust environment with higher rainfall, in which vegetation regrows quickly. Much of northern and western Europe is still able to support productive intensive agriculture today, 7,000 years after the arrival of food production. In effect, Europe received its crops, livestock, technology, and writing systems from the Fertile Crescent, which then gradually eliminated itself as a major center of power and innovation.

Standing calf deep in the Euphrates, one can see the outcome of thousands of years of short-sighted, self-interested and then desperate environmental decisions. The modern Turks are now living with the consequences. Though it was clear they lived in a dry land, the successive empires of the Fertile Crescent committed ecological suicide by taking from the land faster than the land could regenerate itself. They discovered too late that the land was far more fragile than they supposed. The devastation moved westward with the political power, just as it is still doing today. The only places that have been able to avoid this fate are those with enough water to permit rapid regrowth of vegetation and with agricultural systems that provide careful intergenerational stewardship of the land. It is exactly as I have been saying since the United Nations proclaimed its International Year of Fresh Water. Our civilization may be fuelled by petroleum and lubricated by oil, but it runs on water. The Turks have learned this and they are spending billions to try to turn back the clock. One only hopes that too much water and too much time haven't flowed past the banks of the Euphrates to bring the land back, and that the "economic miracle" GAP proposes to perform in southeastern Anatolia is indeed possible. It is not going to be easy to make the miracle happen, for much damage has been done. But the Turks have more experience this time around, and better technology. But the land has been left out to dry for a very long time. As huge as water's capacity to heal may be, it is a tired land the Turks have set out to save. But it is the only land they have, and if they can pull this off, they will have a model they can share with the world.

I would have been very happy to have spent the rest of the afternoon wading in the fabled waters of the Euphrates, and I am sure others would have loved to do the same. This was what we had come to Turkey to do. Our meditations on all the history that was flowing around us were interrupted, however, by a young Turkish farmer who charged his skinny horse and the small cart it was pulling down the bank and into the river beside us. As the horse drank deeply and thirstily from the historic waters, other participants in the forum appeared on the bank above to remind us that it was time for us to depart for further observations and discussions upstream.

Though it was no secret that this region of upper Mesopotamia is an ecological basket case, the hope remains that the river can be the source of economic renewal and cultural revitalization for southeastern Anatolia. The Turks are counting on the Euphrates to pull another rabbit out of its remarkable hat, but it isn't agriculture they are counting on this time, though that will play a part. What the Turks are banking on is that the Euphrates will generate enough hydro power to drive industrialization of the entire region. So it was understandable that our hosts were very excited to show us how they were going to do this.

Our next stop on the winding road north from Birecik along the banks of the Euphrates was the town of Halfeti, one of the 35 communities that had to be moved as the Birecik reservoir began to fill. Looking down on this newly constructed town, it seemed a most pleasant place. If you didn't see what you were really looking at, it would look as if Halfeti were a tiny but inviting resort village on the shore of a sparkling lake. On one level, I suppose, that is what it really is, but on another it is a symbol and an embodiment of what it takes to bring into existence the hydraulic society the Turks are trying to create.

Some 6,500 people were displaced in creating the Birecik Dam and its reservoir. Residents affected by the development were offered three options: GAP would move their existing house to a new location; situate them in a new home such as the ones built for displaced locals in Halfeti; or provide cash compensation for those who wanted to move elsewhere. As is inevitably the case in all such expropriations, not everyone was happy with what they got. But the challenges associated with the GAP project go far beyond the dislocation of locals.

While our gracious hosts reiterated the sustainability aspects of the GAP dream, it struck me how difficult it must be to address environmental concerns, which are equivalent to industrial development interests in any truly sustainable development formula, in a landscape that has been so greatly diminished over time. What is troubling is that Turkey may find itself in the old trap of having to do just one more thing – another hydro-development here, another special industrial zone there – before getting around to seriously examining the costly but necessary environmental aspects of the sustainability circle. This is certainly not a problem unique to Turkey. No matter how wealthy nations become, there just never seems to be enough money or resources left after the development "urgencies" are addressed to put enough effort soon enough into environmental monitoring and thresholds. This is particularly so in areas that are already degraded. For this reason, the bare hills above Halfeti are a good place to reflect on the kind of world people will be born into when the Canadian West has been inhabited as long as the Euphrates Valley has.

Though these issues can be addressed over the long term, they are often eclipsed by higher-profile political concerns. Some of these dog the GAP project. If, in fact, the rather unilateral way this development was planned and undertaken has irritated both Syria and Iraq, problems associated with relocating the little town of Halfeti might just be the tip of an iceberg that could turn out to be the mother of all icebergs standing in the way of replumbing the Middle Eastern world.

Jutting quaintly out into the sparking reservoir, Halfeti is a modern iteration of the hydraulic empires that had their origins in this valley ten millennia ago. It is a contemporary expression of the ceaseless and apparently irrepressible human desire to concentrate power in the hands of those who control water. Like all things economic in today's world, Halfeti represents both a text and subtext. The text of this picture-perfect scene is progress and the opportunity to create a better world for the next generations of Turks. The subtext is the sustainability of the project and every other one like it in the world. In pleasant little Halfeti, text, subtext and context appeared to be unreconciled. It seemed to me that if one scratched the surface of this postcard, one would hear all the locals squabbling. Or maybe all those arguments weren't happening at all except in my own mind, where doubt has been growing for years about how sustainable the world we are creating will actually prove to be.

The generosity, attention to details of hospitality, and concern for our security displayed by our Turkish hosts were once again demonstrated when we regained highway E90 after visiting Halfeti. As we had been warned, the highway was bumper to bumper with slow-moving transport trucks, a fact that was much complicated by road construction. The police literally cleared the way for us, using lights and sirens to part the long lines of clogged traffic until we at last arrived at the booming town of Şanlıurfa.

3. ŞANLIURFA

Your every park a fallen Eden spells me so.

— James Lovett, *O Istanbul: Poems
for a Turkish Album*

Like all the cities in southeastern Anatolia, Şanlıurfa has a long and remarkable history. First settled by the Hurri peoples around 3500 BCE, it was occupied in turn by Hittites, Assyrians, Greeks and Romans. After conquering it in about 333 BCE, Alexander the Great named the city Edessa. The name apparently stuck until the Ottoman sultans had a go at empire, which also meant establishing a completely new cartography with revised place names based on their political aims. The city is perhaps most famous for its association with the Biblical prophet Abraham, said to have been born at or near a pool of fresh water that is now Urfa's most important visitor attraction. There is also a cave in the pleasantly landscaped gardens at the foot of the citadel where Abraham was said to have hidden from the vengeful Assyrian King Nimrod. Urfa was also the centre of the Christian Nestorian movement, named for one Nestorius (386–450), who was patriarch of Constantinople from 428 to 431. The city was also the capital of a crusader state in the 11th and 12th centuries, which sounds like a long time ago until one considers that Urfa had already been occupied for about 2,800 years before anyone even thought of a crusade. Urfa earned the prefix Şanlı, which means "glorious," through its resistance to the attempted French invasion in 1920 that led, as mentioned earlier, to the rise of Atatürk and the modern Turkish state. Today, Şanlıurfa is an industrial city whose future is very much tied to the success of the Southeastern Anatolia Project.

It was twilight when we arrived at the Harran Hotel at the busy heart of the city. This property was supposedly the best hotel in Şanlıurfa, and it was definitely a welcome sight at the end of a long day of very intense perceptual engagement and wonderment. As we were to spend two nights here, I intended to take full advantage of the time available to reflect on all we were seeing and learning, but there were too many things underway and far too many interesting conversations going on to allow time for writing. I was beginning to understand why the Rosenberg Forum had been designed in the manner it had. At no time in my life have I been more engaged in where I was travelling and so immersed in the sharing of experience with energized others who I could see had so very much to teach. This was rapidly becoming a life experience upon which I needed to concentrate completely. I decided to live as fully in the moment as possible and then write about it later.

One of the most interesting people I met in Turkey was an Egyptian who happened to be the chairman of the Nile section of his country's Ministry of Water Resources and Irrigation, with an office in Giza, adjacent to Cairo. He and I began talking on the bus about the nature of his work and his perspectives on the future. He saved a place for me next to him at dinner, which we had outside, next to the hotel pool. He is a founding member of a new school of thinking which argues it is completely unproductive for riparian states to go to war over water. His analysis is based on his experience working with the ten states that share the water of the Nile. His logic is compelling and applicable to some extent to any watershed. For him, integrated watershed management is all about drainage.

Water, he reminded me, flows downhill, or at least it should. "Think about Holland," he said. "If the water that fell as rain in Holland wasn't quickly drained away, the country would quickly be submerged. The most efficient drainage mechanism is one that uses the water and then lets it go downstream. Some countries still think of water as something they own. They think that when water drains away from them, they are losing something. But in places of substantial rainfall, the water cannot be kept. One way or another it drains away. So downstream countries are going to get that water eventually, even if upstream countries do not co-operate in its management. Thus there are two ways to look at water issues. You can create conflicts that cost economically and politically, or you can recognize that benefits and costs can be shared through co-operation."

It is this simple, logical thinking that this senior water policy expert and his colleagues have been working to put into relief among countries that share the Nile. No matter which country you live in, the same unit of water can supply power, keep fish alive, irrigate fields and provide transportation and recreation. Each of these activities is a benefit, each has a value. Drainage, in this context, is a cascading economic benefit. To enjoy this benefit, water has to flow. By extending this kind of rudimentary logic, it should be possible to avoid serious conflict and make war, at least over water, unnecessary.

This pleasant Egyptian's optimism was contagious. With trust, he claimed, it is entirely possible to overcome a 5,000-year history of conflict over water. He knows this because the ten countries that manage the Nile are beginning to make this happen. This in itself has considerable importance as a model of practical optimism. What this seasoned expert is suggesting is that you *can* leave 250 generations of historical baggage at the door and progress to a new future. All you need is trust.

The first stop in our field trip around Şanlıurfa was the ruins of the ancient city of Harran. Established in 1900 BCE, it is an important place in Biblical history, for it too is associated with Abraham, who is a founding prophet in three of the world's great religions: Christianity, Judaism and Islam. You might say Abraham was one of the Big Three. Perhaps even more interestingly, Harran is held in some circles to have been part of the original site of the Garden of Eden. Though such a claim would be difficult if not impossible to substantiate, the legend has a great deal of charm and no small amount of symbolic significance with respect to our forum on avoiding conflicts in transboundary water issues. The Garden of Eden had to have been somewhere, so why couldn't it have been here, just off highway E99 south of Şanlıurfa, in the desert of all our lost possibilities, just a few kilometres from Akçakale on the Turkish border with Syria?

Though it is hardly appropriate to joke about it given how close we were to the Syrian and Iraqi borders, the amount of security accompanying our party might lead one to think we were trying to bust into the Garden of Eden. While we had become comfortably used to being accompanied by police, there were more than usual today, probably because we had important politicians with us. As we convoyed south, two police cars went ahead of us and a police car and an ambulance followed immediately behind. While we felt very comfortable, clearly the police were being extremely careful. Though they had been quite helpful in clearing traffic on the main road between Gaziantep and Şanlıurfa, there was little need for this on the virtually empty road south to Harran. What concerned them was something, perhaps a bomb or rocket attack, that could injure a large number of people. That apparently was why the ambulance accompanied us.

At Harran, not even the local children selling trinkets were allowed to approach us, at least at first. Everyone appeared more relaxed after we spent some time wandering through the main ruins. Perhaps we were too absorbed in our own affairs to notice, but nowhere on this trip did we have cause to feel any sense of hostility toward us, or any hostility at all. We didn't feel it in the streets of Istanbul, Gaziantep or Şanlıurfa.

The country appears calm, even though around it are forces that are currently positioned to tear the whole world apart.

Harran's association with Abraham and the Garden of Eden makes it a compelling place to visit. As Jared Diamond pointed out in *Guns, Germs, and Steel*, Mesopotamia is a very good place to visit if you want to contemplate human influence on landscapes over a very long period of occupation. It is in some ways regrettable that important literary accounts such as the Bible focus on character and plot at the expense of details of setting that could be so useful to us today. But even with limited available comparative information, we can get an idea of the transformation of this area over the past 4,000 years. If indeed Harran was the site of Eden, then we can begin with Biblical descriptions of what that environment was like and compare it to what paleobotany tells us about the region at the time of earliest human settlement. To complete the analysis, all we have to do is look around at what we see before us today, and the long-term human impacts should become quite obvious.

The present-day Harran plain is dry and hot. The soil appears of limited productivity and the vegetation is sparse. The temperature while we were there was approximately 35°C. It looked to me like a desolate place to live. That is the Harran of today. The question, then, is whether this is different from what existed in the past. How was the Garden of Eden different ecologically from what Harran Plain is like now? Though details are lacking, both the location and the nature of the Garden of Eden are described in Genesis. Here, with some personal inter-·pretation, is what the Bible says about the world's most famous garden paradise:

Genesis 2

2.1 *Thus the heavens and the earth, and all the host of them were finished.*

The Big Bang had banged.

2.2 *And on the seventh day God ended His work which He had done, and He rested on the seventh day from all His work which he had done.*

In this passage it appears that God would have been much helped by having an editor.

2.3 *Then God blessed the seventh day and sanctified it, because in it He rested from all His work which God had created and made.*

Ah, those lucky sevens!

2.4 *This is the history of the heavens and the earth when they were created, in the day that the Lord God made the earth and the heavens,*

This is a pretty sketchy as histories go; two brief opening passages – a slender Genesis.

2.5 *before any plant of the field was in the earth and before any herb of the field had grown. For the Lord God had not caused it to rain on the earth, and there was no man to till the ground;*

This isn't entirely consistent with the opening of Genesis, for it must have rained a very long time to create the oceans.

2.6 *but a mist went up from the earth and watered the whole face of the ground.*

Okay, maybe evaporation from the oceans created the rain.

2.7 *And the Lord God formed Man of the dust of the ground, and breathed into his nostrils the breath of life; and man became a living being.*

Are we to surmise from this that the dust used to create Adam was the fine powder we find on the ground today at Harran?

LIFE IN GOD'S GARDEN

2.8 *The Lord God planted a garden eastward in Eden, and there He put the man whom He had formed.*

Note that the garden was east *in* Eden, not East *of* Eden, where novelist John Steinbeck had us end up.

2.9 *And out of the ground the Lord God made every tree grow that is pleasant to the sight and good for food. The tree of life was also in the midst of the garden, and the tree of knowledge of good and evil.*

God made in Eden the mother of all great mixed-wood forests, and a tree of life. Even though Eve ate the fruit of this tree, we still don't know the damn difference between right and wrong.

2.10 *Now a river went out of Eden to water the garden, and from there it parted and became four riverheads.*

There were four great rivers in Eden.

2.11 *The name of the first is Pishon; it is the one which encompasses the whole land of Havilah, where there is gold.*

Ever notice there is only one letter's difference between "god" and "gold"?

2.12 *And that gold of that land is good. Bdellium and the onyx stone are there.*

God was something of a prospector.

2.13 *The name of the second river is Gihon; it is the one which encompasses the whole land of Cush.*

The land of Cush is now Ethiopia. Is Gihon the upper Nile?

2.14 *The name of the third river is Hiddekel; it is the one which goes toward the east of Assyria. The fourth river is the Euphrates.*

The Hiddekel is also known as the Tigris. We are in Eden.

2.15 *Then the Lord God took the man and put him in the Garden of Eden to tend and keep it.*

It looks like Adam waded in the Euphrates just like me.

2.16 *And the Lord God commanded the man, saying, "Of every tree in the garden, you may freely eat,*

Okay, the trees in the greatest mixed-wood forest in the history of the world also bore delicious fruit.

2.17 *"but of the tree of knowledge of good and evil you shall not eat, for in the day that you eat of it you shall surely die."*

Stay away from that tree, Adam. The knowledge of good and evil will poison you – oh yes it will!

2.18 *And the Lord God said, "It is not good that man should be alone; I will make him a helper comparable to him."*	All of nature abhors a vacuum, especially a lonely one.
2.19 *Out of the ground the Lord God formed every beast of the field and every bird of the air, and brought them to Adam to see what he would call them. And whatever Adam called each living creature, that was its name.*	The message here, Adam, is that amidst the great forests, the wild animals and the amazing birds, you will never be hungry or alone, but to ensure that, you have to careful not to undermine creation.

This is what the Bible says about the Garden of Eden before going on to the stories about Adam and Eve, the snake and the apple, and expulsion from Paradise. So what does this description say about Eden, and how was its pristine ecology altered after the Fall?

The first thing we learn is that the garden was only part of Eden. The garden was "eastward *in* Eden," which suggests that areas surrounding it were not the same ecologically. This is consistent with what we know Mesopotamia was like in the past, or at least far enough back to embrace the Eden legend. The Fertile Crescent had a dry heart completely surrounded by spectacular forests. These forests were substantial enough to make the region humid.

According to Genesis, God created in Eden what we can imagine as one of the most beautiful and diverse mixed-wood forests ever to have existed in the history of the world. If one wanted to take Genesis literally, there was also a single tree, the Tree of Life, which symbolized the cumulative wonder that was the glory of Eden. I imagine the Tree of Life not as a tree that actually existed but as an overarching sense of dynamic place that must have pervaded Eden before the Fall. It was the sum of all that mixed-wood diversity, which in terms we can understand today would have placed Adam comfortably in the midst of a self-willed, self-sustaining ecosystem, a garden no more in need of tending than a coastal rainforest. Adam, the quintessential hunter–gatherer, lived in paradise.

It was evidently unfortunate, then, that Adam's wife began having long philosophical discussions with perhaps disreputable but clearly articulate snakes. It is interesting, though, that the Fall was not principally moral or spiritual, at least at first. The Fall, as the Big Guy had hinted, was ecological. It was only after the ecology of Eden went into decline that the world went to hell in a moral and spiritual handbasket. Here is how Evan Eisenberg describes it in his book *The Ecology of Eden*, casting the symbolism of the garden in ecological terms and the Fall of Man in terms of deteriorating environmental integrity:

> There are certain places on earth that play a central role in the flow of energy and the cycling of water and nutrients, as well as in the maintenance of genetic diversity and its spread by means of gene flow. Such places provide many of the services that keep the ecosystems around them (and the biosphere as a whole) more or less healthy for humans and other life forms. They help control flooding and soil erosion. They provide fresh infusions of pollinating birds and insects, which grant continuing life to many of the plants (both wild and cultivated) we

take for granted. They regulate the mix of oxygen, carbon dioxide, water vapor and other ingredients in the air and keep its temperature in bounds. They are spigots for the circulation of wildness through places made hard and almost impermeable by long human use.

Eisenberg is proposing that Eden was the place on Earth where nature provided the greatest range of environmental benefits and services completely for free. With each diminution of benefits or reduction in the range of services it provides to humans and the rest of nature, a given ecosystem becomes less and less like Eden.

In this context a mere bite into an apple precipitates a relationship change with the world caused not by a moral crisis but by the fact that the world is diminished and therefore less able to provide as completely for us as it did in the past. Our relationship is diminished because our partner in the relationship – the natural world itself – is only a shadow of what it formerly was. We are less because she is less. As so often happens in relationship declines that make the world a measurably poorer place, the process of diminishment and loss can be a painfully gradual one. Eisenberg:

> The Fall of Man is not a sheer drop. It is more like a scrambling outward and downward that must feel at times like a helpless sliding, as when your heels refuse to catch in a scree slope, at times like a giddy romp, at times, even, like a triumphant ascent. The waves of human-led change – our alliances with certain species, and roughshod crushing of others; our unhousing of the soil community, our stirring up of

long-buried hydrocarbons – are a matter of concern to creatures on whom sermons about free will might well be lost. In various senses, we take a lot of non-human nature down with us.

In other words, the damage to the global ecosystem we have committed during our Fall may bring about the fall of the world itself. If you were another species, you may not want to be roped to us on the mountain.

But even in barren Harran there is evidence of important Turkish efforts to slow and perhaps even reverse the Fall. To bring back Eden, you have to bring back the Euphrates, or at least its powers. The hope is that here, in the home of Adam, the first gardener in the world, irrigation can be made to revive an agriculture that will hopefully restore the land to some approximation of Eden. That, at least, appears to be the plan. It will not be easy. Evan Eisenberg speaks to the challenge of restoring the world to its primal glory:

> If we want to dream of Eden, we had better go back to sleep. But if we want to understand the dream, we had better learn something of the past from which it sprang. Only then can we hope to divine its meaning for the hard daylight hours ahead.

After visiting the ruins of Harran, we visited a national historic site, or equivalent, which featured the remarkable cone-roofed, or beehive, mud houses that are part of the traditional architecture of this hot, dry area of southeastern Turkey. These structures are designed like row housing with low doors. We entered what we thought was the living room only to find it connected to all the other rooms laid out in a long line under a dozen cone-shaped roofs.

Each of the rooms was surprisingly well furnished, with beautiful Turkish carpets on the floors and walls. What was truly amazing was how cool it was in each of the honeycomb rooms. By the time we reached this site, it had to be close to 40°C outside in the sun, but these quarters were a cool and shady 25°C.

It was clear from looking around the town that had sprouted up here at the ruins of Harran that this form of architecture was much favoured in this desert place. Some of the nicer units even had satellite dishes on the roofs. There was also a hotel composed of these unusually shaped little cells, where I wouldn't have hesitated a moment to stay. The only problem that could cause any inconvenience in these structures, I surmised, might be snakes and desert scorpions. As the doors led to below-grade accommodation where locals and visitors slept, it wouldn't be at all surprising if critters occasionally crawled in to share warmth on cold desert nights.

I would have liked to have investigated this more carefully, but in the entire time I spent in Turkey I never came across any reference material on the country's natural history in either English or Turkish. It was almost as if the country didn't have a natural history. The popular guidebook I had with me mentioned snakes and scorpions but did not identify species.

After visiting the cone-house village, we drove a short distance out of Harran to see the ruins of the first Muslim university. As we walked from the bus to the viewpoint overlooking the ruins of a central tower amid a cluster of stone buildings, we found ourselves ankle-deep in dust as fine as talcum powder. This is what the soil of Eden has become, and it is this soil that would have to be somehow restabilized and then restored if irrigation of the Harran plain is to succeed. It looked impossible, but in thinking this I realized I was underestimating the soil-restoring power of water.

In *Rivers of Empire*, Donald Worster tells a story about what parts of California were like before irrigation transformed its Mediterranean-zone deserts. The description, particularly of the dust, is so similar to what we experienced in Harran that it bears mention and consideration:

> John Woodhouse Audubon, the younger son of the famous bird painter, set out from New York City in 1849 for the California gold fields. Landing on the Texas coast, he and his party struck out across the Southwest to San Diego, following the Gila River for part of that distance, then trudging through the desert now called Imperial Valley. "Our road," he wrote, "is garnished almost every league with dead cattle, horses or oxen; and wagons, log chains, and many valuable things are left at almost every camping ground by the travelers; we ourselves have had to do the same, to relieve our worn and jaded mules, able now to carry only about a hundred pounds." Dust flowed over their shoe-tops, rose in clouds to fill their eyes, choked animals and men alike. For long stretches the only life they saw was creosote bushes, sunflowers, and a lone vulture. Ordeal, suffering, bones, death – nothing the desert inflicted could induce them to tarry awhile.

If irrigation could turn the desert Audubon travelled through into the Imperial Valley, which is currently one of the richest agricultural regions in the world, then perhaps the dust of the Harran plain could be similarly redeemed. I hope the Turks can pull it off and then make their success stick. The ruins of the first Muslim university are a stark reminder of what can happen if you fail.

Soil types notwithstanding, it would appear that with proper irrigation it may very well be possible to restore much of the Harran plain to high agricultural productivity. Certainly the Turks believe this and are working to make it happen. A number of forum participants, however, appeared to have concerns.

I thought about these questions as we drove away from the ruins of the 4,000-year-old university. If this institution had survived, we might have been able to go there to learn all about what we lost when we destroyed Eden. On the other hand, if a university were still in existence at Harran, it would likely mean Eden had not been completely destroyed. Modern-day Turks would have been relieved of much the burden of trying to recreate a semblance of their lost world, and the Rosenberg Forum might be working on matters of water policy that far transcended avoiding conflict in transboundary water issues.

While our driver steered toward Atatürk Dam, I had the pleasure of sitting next to a Canadian who at that time was executive director of an organization created to preserve and protect the Danube River. He offered a number of very interesting and immediately applicable ideas for public engagement with our own watersheds here in Canada. The first of these was what he called "Danube Day," an annual celebration in the 18 countries through which the river flows. In addition to dozens of events held in communities along the way, ports show their solidarity by sounding their sirens in sequence to celebrate how each of the 18 countries is linked by Europe's most important waterway. Another outstanding example was a program his organization created in public schools in association with the annual Danube celebration: a competition to identify an annual

"Danube Art Master," the student whose work best portrayed the spirit and nature of Europe's great river.

Beyond our discussion of public programming, my interlocutor also offered some thoughts on the mechanisms of potential co-operation among sovereign nations sharing the Danube. His most important observation was that it was very difficult to change the fundamental legislation that frames international law concerning how this watercourse should be managed. This would become a theme throughout the forum. Legislation for managing water resources is, in many instances, outdated and inadequate to the challenges facing integrated watershed management almost everywhere in the world. In this regard it could be said that existing legislation, or perhaps more accurately the inability to readily change such legislation, is an "infrastructure trap" in its own right, comparable to the more concrete infrastructure traps we have created through built form. This expert posited that efforts to modify legislation would be wasted, but real gains could be made if he could find ways for the 18 Danube countries to work voluntarily and collaboratively toward more efficient overall watershed management, with the existing legislation functioning as a backstop.

Events had demonstrated that a river can be a source of common interest that transcends historical conflicts between nations. Even after the terrible Bosnian hostilities, the Danube River proved to be the first and best way for war-torn countries to begin to relate to one another normally again. The river formed a useful avenue for finding ways to co-operate, with the aim of ending conflict rather than stoking it, and establishing regional peace.

What was being done along the Danube was exactly what we were proposing for Western Canadian watersheds. We don't have time to change legislation. The only option we have left in the short term is to use legislation as a foundation, then transcend it through creative co-operation in the hope that such collaborative efforts will improve the vitality of our declining watersheds.

There is greater urgency in this than most people realize. Some of the rivers that flow from the eastern slopes of the Rockies have been reduced by climate change and other factors to 40 per cent of their historical flows. We can't wait for legislators to act. Our future success in managing water and land use issues in the Canadian West will depend on our ability to work together to craft watershed and other forms of compacts that leave rights in place while respecting larger ecosystem and society goals. These compacts can then be shored up with legislation as their durability and sustainability become proven.

The Atatürk Dam is the crown jewel of the Southeastern Anatolia Project. The moment we arrived we were invited into an elegant, air-conditioned theatre with plush blue seats to hear a presentation by a vice-president of the GAP initiative who put forward an impressive array of statistics in support of the project's goals. His main point was that this massive state project was successfully improving land use while increasing human capacity in order to prove sustainable, integrated social and economic development on a regional scale. To accomplish these goals, the government water authority, the General Directorate of State Hydraulic Works (in Turkish, Devlet Su İşleri, or DSİ), had effectively replumbed southeastern Turkey.

There are 22 dams on the Euphrates (the Firat basin, as the Turks call it) and the Tigris (Dicle) rivers. Nineteen of these dams also generate power. The objective is to put 1.7 million hectares into irrigation and, as noted earlier, to generate some

27 billion kilowatt hours of electricity annually. The project embraces significant improvements in almost all aspects of regional life, including agriculture, industry, transportation, communication, urban and rural development, health, education and tourism.

The ultimate goal of the Southeastern Anatolia Project is a staggering 445 per cent increase in the region's gross productivity, a 200 per cent increase in its per capita income, and jobs for 3.8 million people. As befits such bold objectives, this was not an inexpensive project. The total cost of GAP at the time was expected to be US$32-billion.

The vice-president's statistical presentation was dizzying, which I am sure was its intent. Turkey was clearly planning utilization of the nation's rivers on a Chinese scale. The Tigris–Euphrates system, which drains 28 per cent of Turkey's surface water, is about to resemble the Columbia, the world's most damned river, so to speak.

The state water authority is responsible for four functions in this great harnessing of the life-giving rivers of the former Fertile Crescent. A huge organization with sweeping powers, the DSİ is responsible for planning, coordination, monitoring and evaluating the success of this grand effort to modernize Turkey and, among other things, bring it into the European Union.

In support of claims of sustainability, the agency was careful to profile social as well as economic development efforts. The vice-president and his colleagues showed images of one of the 78 community centres they had created for women. Some 85,000 women had taken literacy and health-training courses at these centres. Similar support was being offered to dryland farmers, for whom demonstration orchards had been created and some 176,000 pistachio trees planted. These are astounding programs.

The conference was going along swimmingly until one of the forum participants raised her hand and posed what at first appeared to be an innocuous and quite legitimate question. "When you hit 100 per cent of your irrigation goals," she asked, "what will happen to countries downstream in terms of water supply?" There was a momentary silence in the room, which the participant, a Canadian water policy expert, calmly decided to fill with some clarification of her point. "It does not appear," she said, "that your irrigation plans have been accepted by either of your downstream neighbours." With that simple remark the tone of the Turkish presentation instantly changed. The speakers and translators became very agitated and the dialogue faltered. When those on the rostrum could not arrive at any consensus over how her question should be answered, an eloquent Turkish diplomat stood up and saved the day for his countrymen. Though he did not exactly answer the expert's question, he certainly put into relief Turkish foreign policy with respect to its legal transboundary riparian rights.

The diplomat was not the kind of person who makes mistakes in legal interpretation. As forceful as he was in his presentation, however, and as respectful as we were of his obvious expertise, not everyone in the audience seemed satisfied that the question posed could be properly answered by referring solely to legal precedent established by a controversial 1946 treaty that might no longer be relevant to the transboundary circumstances or applicable to the political context of the region today. Though it evoked tension, this brief debate was crucially important to the ongoing process of establishing dialogue that could get to the heart of difficult transboundary water management matters, not just in Turkey but in every country represented at the forum. To pose questions about issues like these was the reason why we came to Turkey. Change the name of the countries or jurisdictions involved, and the problems we were discussing here could be seen as almost universal. Was it practical for Turkey, or any other country, to make unilateral decisions on large-scale water projects based solely on the legal strength of outdated treaties? Would the decisions Turkey was making with respect to the Tigris and Euphrates ultimately sustain economic development throughout the larger region through which these rivers flow, or would such benefits be confined to Turkish borders? Were the country's plans truly sustainable, or would they result in transboundary water conflicts in the future? Could tensions between upstream Turkey and downstream Syria and Iraq be avoided as Turkey moved to maximize its own irrigation potential? If one were to replace the still highly controversial 60-year-old agreement with something new, what might be done to ensure that the new arrangement would be fair, equitable, durable and adaptable in cases where climate change and other impacts might break all the rules?

These very same questions could be asked where I live, as the province of Alberta may be using potentially outmoded legislation as a means of determining downstream water availability on major rivers that flow to the other prairie provinces. The very same questions were being asked at the time and continue to be asked in the Columbia basin, where residents were beginning to plan for possible reconsideration of the Columbia River Treaty between the Canada and the u.s. There likely wasn't a country represented at this forum that didn't have a similar situation already brewing or about to be.

Though the issue of fair, equitable and durable treaties did not surface directly, the Turks did concede that the current situation was packed with potentially contentious outcomes. The Turkish delegation pointed out that agreements would eventually be created by a joint commission expressly constituted to address these issues. In the meantime, however, the resettlement of 6,500 people would continue, and dam and irrigation development would proceed on the timetable established by the DSİ for the GAP initiative. The GAP plan had to go ahead.

The debate left many participants with more questions than answers. I wondered, for example, if American foreign policy might be influencing the official Turkish stance toward the potential needs of its downstream neighbours. The Americans had been very clear that they considered Syria to be as great a threat as they had considered Iraq. They had made very bold statements about their willingness to use the invasion of Iraq as a first step to crushing hostile political regimes such as exist in Syria and other Islamic countries in the region. Clearly the funding for the GAP project would not have been forthcoming without a great deal of American support. One would be very surprised if current and future U.S. administrations wouldn't have at least something to say about how the redevelopment of southeastern Anatolia was being undertaken, at least in terms of its contribution to the political stability of the region. We were left with other questions too concerning sustainability.

One of the speakers noted that drainage water from the Harran plain flows into Syria. The GAP plan included the construction of two wastewater treatment plants at the Syrian border to prevent agricultural pollutants and excess nutrients generated through irrigation from causing water quality problems downstream. Then one of the GAP engineers dropped a bomb: "The

Harran plain is shaped like a bowl. It doesn't have good field drainage. There is fear of salinization."

Sustainable irrigation depends on good drainage. Without it, salts accumulate in the soil and make it useless for agriculture. Once soil is salinized it can take 500 years to reverse the damage. Though it was not too late, perhaps, to address the drainage issues in new irrigation lands on the Harran plain, this disclosure put into clear relief how difficult it is to make every aspect of megaprojects of this sort work. Preventing salinization is the first principle of sustainable irrigation agriculture. If salinization cannot be prevented, no amount of money spent on dams and canals will help you. The economic benefit of projects in which salinization cannot be controlled or reversed will be confined to the period in which a large labour force is employed in construction of dams and reservoirs. If salinization cannot be prevented, then what is left of Eden could be under final threat. Despite investment in knowledge and resources and billions of dollars in aid, the ecosystem that was Eden will have gone from forested plains to semi-arid and scrub desert to salt plain wasteland in little more than 10,000 years. This could hardly count as a promising omen for long-term human settlement here or anywhere else on this planet. If the Turks failed, southeastern Anatolia could sound an alarm for our entire civilization.

After the presentation, we were invited out the front doors of the theatre to behold the magnificent arc of the Atatürk Dam, the sixth-largest structure of its kind in the world. Its sheer scale took my breath away.

We then boarded a bus to go out onto the dam itself, behind which is an 870-square-kilometre reservoir. How utterly alien it was too see what at first appears to be a huge natural lake in the middle of a desert. Here amidst the parched hills is all the water in the world. Look way down and you see tiny human figures swimming through a blue eternity of perfectly clear water.

Though it was a Sunday and the local swim club could have quite logically been practising in the lake, I suspected the swimmers were performing just for us, as was the small sailboat tacking back and forth on the reservoir beneath us. I found the view eerie. There was no water elsewhere in south-eastern Anatolia, I surmised, because it was all in this reservoir. How could a land be so dry when it had so much water? And what was happening elsewhere, if all this water was here?

Back in Şanlıurfa we were to be dinner guests of the DSİ. A bus took us from our hotel to one of their office complexes, where dinner was served elegantly and efficiently while conversations among forum participants continued unabated. The energy of the entire group had clearly risen as a result of the discussions at Atatürk Dam.

When we returned to the hotel, there was a huge Turkish wedding in full swing in the pool area where we had dined the previous night. As the hotel had no public space where our discussions could continue, I decided it was a good time to try to give meaning and value to this rich and very full day by sleeping on it.

I was very amused in passing through the crowded foyer to notice a number of my fellow Rosenberg colleagues out on the floor, dancing to loud Turkish music amidst the press of the wedding guests. It was a very civilized wedding, at least by Canadian standards. My room vibrated with the music until just after midnight, when the guests went quietly home.

4. ANKARA

The power of water over history is a very old
discovery. The earliest map of the Middle East,
dating from the eighteenth century B.C., shows
the River Euphrates dividing the lands of the earth
into two islands. Around the perimeter of the map
flows the vast circling sea, Oceanus, the ultimate
river from which all lesser rivers come, at once the
source and the destination of Euphrates. Human
existence must be carried on, the map indicates,
within that watery loop and along its pathways.

— Donald Worster, *Rivers of Empire*

With the preconference field trip over, our group was scheduled
to travel from Şanlıurfa to Ankara on a flight that only departed
three times a week. Henry Vaux had made it clear that if you
missed this flight, you'd miss the forum. As a result, everyone
arrived early for the bus. I was surprised that the phalanx of
police cars and the ambulance continued to accompany us even
on the 15-minute trip to the airport.

As it turned out, I had a row to myself on the flight to Ankara
and happened to look down just as we passed over the Birecik
Dam and its reservoir. To the north I could also see the reservoir we had visited at the Atatürk Dam. If anything, these big
bodies of water looked even more alien than the Atatürk reservoir had appeared from the ground. From the air, the GAP project seemed every bit as monumental as it had been described
to us in terms of its overall influence on southern Turkey. There
can be no question that the DSI was having a major influence
on the Euphrates River. From a mile above the ground, it was
possible to see how the run of the river had been contained,

first by one dam and then by another. The Euphrates was just a series of slack-water lakes like the Columbia. No wonder there was no current when I stood in the river at Birecik. The water doesn't move unless it is ordered to.

I would have felt considerably better about what I was observing if previous empires that had overseen water diversions on scales not even close to what I could see from the plane had not in the end created the desert through which the Euphrates now flowed. Looking down on this most historic of all rivers, I realize that the Saskatchewan is western Canada's Euphrates and Nile, and parts of our plundered province are beginning to look far too much like the burned-off hills and plains of the once-Fertile Crescent, the cradle of civilization and the province of our former Eden. I fear for my Euphrates. Once again I am guided in my observations by Donald Worster and his deliberations on human control of water, particularly for irrigation:

> Control over water has again and again provided an effective means of consolidating power within human groups – led, that is, to the assertion by some people over others. Sometimes that outcome was unforeseen, a result no one really sought but dire necessity seemed to require. In other places and times, the concentration of power within human society that comes from controlling water was a deliberate goal of ambitious individuals, one they pursued even in the face of protest and resistance.
>
> … Consequently, nothing suggests more clearly than the study of irrigation in history how dependent societies may become, not merely on water, but on their manipulation of its flow. And nothing makes more

clear the link between water control and the social orders humans have created than irrigation history.

And it is that history that concerns me now, whether I am flying over the Fertile Crescent of Asia Minor or the Great Plains of North America. The most salient point this landscape makes on its own behalf is that humans can only survive over the long term if they think and act with the long term in mind. Our technology will not save us if we suffer a failure of restraint. Despite all our knowledge and wealth, our future still hinges on fundamental human traits, the same traits that have dogged us through all of history, the traits that define us as individuals and as a species. If we want to know the world, we have to understand water, but first we have to understand ourselves.

As Ankara is higher in altitude than southeastern Anatolia, it was cooler here at this time of year than it was in sweltering Şanlıurfa. It was, in fact, most pleasant. Two buses awaited us at Ankara's modern air terminal and we were soon on our way through what was clearly a very cosmopolitan capital city, bound for the Bilkent Hotel, 40 minutes away. I was surprised to observe that we had yet another police escort, this time with sirens, though without an ambulance.

5. THE ROSENBERG INTERNATIONAL FORUM ON WATER POLICY

In the case of technological innovations and political institutions as well, most societies acquire much more from other societies than they invent themselves.

— Jared Diamond, *Guns, Germs, and Steel*

With the inaugural speeches completed, Dr. Vaux calmly

and masterfully raised the level of excitement in the room by introducing the first keynote speaker, a man I had very much looked forward to hearing. If there was anyone in Turkey who understood how important water is to the Middle East and to the world, it was H.E. Süleyman Demirel. Henry Vaux carefully chronicled this man's amazing life, a life very much shaped by a profound relationship with water, as it turns out.

Süleyman Demirel began his professional life as a hydraulic engineer. At 30 years of age, this very bright and energetic man was director-general of all dam development in Turkey. He was the chief, in other words, of the State Hydraulic Works, or DSİ. In 1965 Demirel became Turkey's 12th prime minister. A later military coup resulted in his being banned from politics for ten years. After seven years of persistence and struggle, he managed to force a referendum reversing the ban. Between 1993 and 2004, Süleyman Demirel served as the ninth president of Turkey.

Demirel was a short, stout, balding little man you might not notice on the street, but the moment he stood at the podium, he became a sparkling-eyed giant. His subject was close to his heart: the importance of transboundary water management. In his speech he said some very important things, of course, but one of his statements could ultimately have import in my own life. While it would serve little purpose to repeat what already exists in print from his speech, one section is important enough to copy in full:

> Why have you arrived in Turkey from all over the world to discuss water management? The answer is quite simple: Water is essential for our survival. Hence the dictum "water is life." It is very true that water is one of the engines of sustainable development. We

cannot fight poverty and hunger without utilizing our water resources. According to the World Water Development Report, prepared by a UN World Water Assessment Program, by 2015 nearly 40 per cent of the world population is expected to live in water-stressed countries. It has also been reported that the demand on water resources will continue increasing during the decades to come. The Middle East, Africa, China and Central Asia in particular will face serious water shortage.

The targets set by the Millennium Declaration of 2000 and the World Summit on Sustainable Development of 2002 reflect the willingness of the international community to address the matter at a global level. In recognition of the importance of the matter, the UN General Assembly took an additional step of proclaiming the year 2003 as the "international Year of Fresh Water." The UN General Assembly also proclaimed the period from 2005–2015 as the international decade for action "Water for Life." These set targets and decisions attest to the importance of managing water resources properly.

It was amazing that I had had to come to Turkey to learn from its former president something so potentially important to my work back home. But there it was: the path I could choose to take at that very moment would occupy me for the next ten years of my professional life. The trailhead – fully identified, compliments of former president Demirel himself – was right before me. All I had to do is walk down it.

Quite unknowingly, the UN had created a decade-long water initiative that just happened to have the same name as a major water strategy that had just been completed by the province of Alberta. "Water for Life" had become a water resource management theme not just for western Canada but for the whole world.

This coincidence aside, back at the Rosenberg Demirel was reminding his audience that the Middle East had historically been a water-stressed region. He noted a remark by the late Israeli leader Yitzhak Rabin about serious water shortages in the Jordan River basin: "If we solve every other problem of the Middle East but do not satisfactorily resolve the water problem, our region will explode. Peace will not be possible." President Demirel disagreed. Exploring what would later become a persistent theme throughout the forum, he argued that management of water resources was one the few things that, if only by sheer necessity, had to bring people together rather than divide them. On this issue Demirel was insistent:

> I agree that the issue of water in our region is complex, emotional and political and therefore it will remain with us in the decades to come. However, I would like to challenge Mr. Rabin's statement, as I do believe that the countries of the region could use water as a tool for co-operation rather than a source of conflict. The states of the region will eventually come to understand that co-operation is imperative for the improvement of the quality of life of their people, socio-economic development and stability in the region. In short, the cliché that the next war will be fought over water has, in my view, become obsolete.

Demirel believed, as many of the other forum participants did, that resolution of water issues cannot occur without building mutual trust and confidence among riparian states. "Parties concerned," Demirel observed, "should first and foremost free themselves from nationalist rhetoric, clichés, emotions and prejudices. Necessary steps should be taken in order to dispel mistrust and create the appropriate environment for meaningful co-operation."

Water will be the vehicle for doing this, for in matters related to water, as Demirel pointed out, we have no choice but to work together. What we are witnessing is nothing less than what Daniel Kemmis reported in his book *This Sovereign Land*: the emergence of watersheds as nested elements in a new political order.

More than all the legal arguments that followed, Demirel's point about finding ways to share rather than divide water, and the world, have stuck with me. In 1900, Demirel noted, there were 1 billion people on Earth. A century later, there were 6.5 billion. By 2050 it is expected there may be as many as 9 billion of us crowded on this tiny blue ball. Climate change and environmental decline, Demirel acknowledged, were likely going to affect water availability in a negative way. As a result, the kinds of issues the Rosenberg Forum was created to examine are undoubtedly going to become more pressing. This, at least, is what Demirel put into relief by his assessment of what our future might be like:

> The large number of transboundary rivers on Earth, the size of the population dependent on these rivers, and the complex nature of the various transboundary water disputes have a direct bearing on the quality of life of billions of people as well as peace and stability at a global level.

The very last thing Demirel said seemed almost as if it was aimed very close to my heart:

Lastly, there is a need for enhancing public awareness
with respect to the importance of efficient use of water
resources. The press and media can play an important
role in this regard. It is our obligation to protect the
quality and quantity of water resources for the next
generation.

Such thoughtful restraint and long-term vision as described
by Süleyman Demirel are hard to cultivate and maintain in
business and political life. For that reason, it would be easy to
be doubtful that conflict over water or any other resource will
necessarily become obsolete in the face of an emerging co-operative imperative. If it is to happen, and Süleyman Demirel was
correct in arguing that it must, then we are faced with an urgent
need to reassess the effectiveness of our most fundamental institutions. Our deepest desires have to be examined and altered.
But that, once again, was the very purpose of this forum.

One of the next speakers about co-operation in managing
transboundary water resources worked for the World Bank.
Also a member of the Rosenberg Forum advisory committee,
he made a number of interesting points that very much supported what Süleyman Demirel had said about the obsolete
nature of conflict over water. Proper transboundary management, he confirmed, was crucially tied to effective co-operation.
While co-operation between countries was important, so was
co-operation between disciplines. He then demonstrated very
concretely why the potential for interdisciplinary and intergenerational co-operation is being recognized internationally.
He first quoted one of his World Bank colleagues to the effect
that "the wars of the next century will be over water."

In arguing that water issues should not automatically be
seen as a source of conflict, this speaker pointed to statistical

research showing that more treaties are being signed by states over water, and that proportionally these were increasing over incidents of conflict. His point was an important one. This same point had been made by Aaron Wolf at the "Mountains as Water Towers" conference in Banff in 2003 and made again by others in discussion at the Harran Hotel in Şanlıurfa. In the long term it is not very productive to go to war over something as absolutely essential to everyone who lives on this planet as water.

The presentations that followed introduced the notion of sustainable development as defined globally, which I assumed was the definition the Turkish government had applied to its GAP project. Members of the Turkish delegation then did something it took me a great deal of time to appreciate: they actually said out loud something that many people had been barely able to countenance in thought, namely that we still don't know how to make development sustainable.

Sustainable development, in Gro Harlem Brundtland's now classic formulation, is development that meets existing needs without compromising the ability of future generations to meet their needs. Sustainable water resource systems are those designed and managed to fully contribute to the objectives of society, now and in the future, while maintaining their ecological and environmental integrity.

Upon hearing this, I realized that what was happening in Turkey, and in parts of Canada as well, was not sustainable by this definition. We have not yet been able to identify a development process which can be designed and implemented to function in a way that is inherently and perpetually sustainable. A light went on inside my head when I heard the Turks make this remark. While some forum participants had reservations about whether or not Turkey had truly achieved the level of Western countries in terms of their overall recognition of the importance of integrated environmental, social and economic development, we were proven wrong. The Turks are fully modern. Though their culture and language are very different from ours, they are just like us in that they haven't figured out what sustainable development really is either.

As I reflected on this, it occurred to me that these were some of the most important remarks at this forum. How could we condemn Turkey or any other country for failing to achieve or even consider the importance of sustainability when there was no known process that could actually be said to be sustainable? Where I had thought some forum participants were simply dismissing the Turkish project on the basis of its lack of sustainability, I realized that something far more important might be happening. We are recognizing in what the Turks are doing that the model we have created – the very model they are using to emulate us and ultimately become like us – is not yet completely developed. I wondered if having to admit this fact was preventing some of us from directly acknowledging our own failure, or from finding appropriate language to warn the Turks away from some of the terrible mistakes we have made in continuing to support short-term prosperity that could very well be at the cost of our civilization's long-term sustainability. To accept the GAP plan as it was being presented was a hard pill to swallow, but to clearly make that point would have required us to stand up before this international forum and admit we have no more idea of how in practice to achieve real sustainability than they do. Based on the fact that we already have the wealth to which the Turks aspire, how could we ever countenance such a disclosure?

A chill ran up my spine. If major projects being developed on the Euphrates could not be made sustainable, what implications did that have for our rivers here in western Canada, such as the Columbia, the North Saskatchewan, the Bow?

Following the Turkish delegation's presentation, there were further troubling questions from the forum, focused on the problem of soil salinization. In response the Turks indicated that the most serious salinization was occurring on the Şanlıurfa–Harran plain, where drainage posed a major challenge to sustainable irrigation. Data from some 763 observation wells drilled on the plain showed that 6 per cent of the area had a shallow water table, less then 1 metre from surface level, during the peak irrigation season. Studies indicated that 8 per cent of the irrigated land was already exhibiting salinity problems.

When pressed for more information on this, our hosts offered that 6000 hectares had already been modified in ways that improved drainage, and that work was presently underway on another 30,000 hectares to ensure that drainage problems didn't lead to salinization. Another 340,000 hectares, however, required similar attention. Only 1.5 per cent of the potential 400,000 hectares of potentially irrigable land in the remaining GAP project area had been properly cultivated to improve drainage enough to ensure sustainable agriculture. Clearly, there was a big challenge ahead.

It was put forward that there was a very real risk in continuing to grow cotton without proper irrigation technology and knowledge. The Turkish response was astounding, at least to me: "For Europe the focus is ecological," they said, "but in Turkey, water is the focus of survival and the alleviation of poverty."

I had to sit back and think about that remark, for it was fraught with all the complexities associated with proper management of our water and other environmental resources. It would have been very easy to condemn this response and simply write the Turks off as being insensitive to environmental considerations. But it wasn't that simple. Though no one mentioned it, we had made exactly the same decision in North America amid the collapse of the cod fishery on the Atlantic coast. Before judging the Turks, it was important to compare our two situations.

Some of the causes of the Canadian cod fishery collapse had been identified. They included the facts that almost everyone thought the resource was inexhaustible, and that private interests demanded to be served before the real issue was addressed. Federal and provincial governments had warred over jurisdiction, so no organization could take charge. Government departments suppressed information, scientists who offered dissenting views were discredited, and short-term political gain was put ahead of the sustainability of the fishery. An environmental catastrophe became an economic disaster and then a social one as well. The worst thing was that we didn't learn anything from any of it. Five years after a moratorium on fishing was imposed, the stocks still showed no sign of recovering, yet we allowed restricted fishing to continue.

The only thing substantively different about the Turkish situation was that their last environmental catastrophe that became an economic and social disaster happened a very long time ago. Though it was every bit as devastating in its time as the collapse of the cod fishery was in ours, its lessons are beyond the memory of contemporary Turks. What happened to Eden happened to somebody else. There is no need to regard the old myths. There was water to be managed and money to be made, just as, in our own case, there was cod to be caught. I hoped that, over time, the GAP project did not become the second death knell of the upper Mesopotamia region. No one in the world would win if GAP became the last techno-straw that finally insulted what was left of Eden out of production.

It struck me once again how important it is to learn from our history. The Euphrates Valley teaches us that economic prosperity can only be sustainable if it calculates environmental costs over the long term. This calculation is not just a luxury that wealthy people like Europeans or North Americans can afford. It is a calculation that is crucial to survival, especially in fragile landscapes. GAP's irrigation and economic development ambitions were impressive, but it should not be ignored that this project marked the second great irrigation era in the history of this region. Every effort possible had to be made to ensure that this iteration of hydraulic empire would be fundamentally different from the previous one. Unless the drainage issue was solved, the results of GAP might eventually not be any different than what happened the first time around. It becomes a question of human nature.

The ancient Greeks understood the problem of human will. They coined a word that describes the gulf that often exists between what we know needs to be done and what we actually do – the difference between the decision we need to make, if you will, and the decision we actually render. That word is *akrasia*: "knowing what is right and failing to do it." Canadians, by this definition, have a bad case of it. We are akrasiacs. So is just about everybody else. The question this Turkish case study demanded we ask was: How do we cross the bridge from knowing the right decisions to actually making them? We can get over our cultural akrasia by going back to basics. And the most basic of all basics is water. Follow the water. The Turks were trying to do just that, but it was not going to be easy.

The next pillar in Turkey's national water policy would cause heart attacks in many of the Canadian circles in which I travel, because they are opposed in principle to any transboundary export of water. Turkey was aiming to fully develop its potential to transfer water from its rivers to neighbouring water-stressed countries.

The Middle East, and Gulf states in particular, were expected to face serious water shortages in the next 25 years. Turkey appeared to be the only country where fresh water was available for transport to those regions, whether by pipeline or tanker. This would be done by tapping the unused waters of rivers that flow into the Mediterranean. To this end, Turkey had built a water treatment plant at Manavgat, on the south coast, where the Manavgat River flows from Beyşehir Lake through Beyşehir National Park and enters the Mediterranean east of the city of Antalya. The Manavgat facilities were capable of exporting 180 million cubic metres of water every year, and could easily be expanded. An agreement between Turkey and Israel had been signed that would allow water to be shipped from Turkey in purpose-built tankers for a period of 20 years. Both countries cited this agreement as the key to easing the Jordan River water crisis. Turkey was further considering using the Manavgat plant to supply water to other Mediterranean countries suffering from water shortages.

The final presenter in the Tigris–Euphrates case study, a professor at Bilkent University at the time, put into relief the great differences that can exist between legal and political expressions of the same problem. Whether the professor intended to or not, he proved through sheer lawyerly assertiveness that legal precedent can be an obstacle in its own right to efforts to co-operate toward integrated transboundary watershed management solutions.

The professor wasted no time on pleasantries. He indicated that the Tigris and Euphrates were the two principal rivers in the Middle East, as well as being the longest, both rising in Turkey before flowing through Syria and Iraq. He bluntly indicated that neither Syria nor Iraq had actual water shortages, despite their misuse of water by employing old-fashioned technologies. He then claimed that some Middle Eastern states that were not even riparian to these rivers nevertheless regarded the Tigris and Euphrates as a panacea for the water problems in Syria and Iraq. This, he charged, was an idea that relied on false facts and an improper conception of the law. The first misconception that had to be addressed, he said, was that Turkey was not a water-rich country, especially when its future water needs were considered. The second misconception, he asserted, concerned international law, which, in principle, aims to settle disputes only among riparian states. In the context of such law, it is up to the state that possesses the water to decide whether to allocate it to some third state that is not an actual riparian, from its own share.

The professor also pointed out something of great importance regarding the standing of such disputes. The disagreements related to the optimum and sustainable use of the waters of the Tigris and Euphrates were considered a legal issue by Turkey, but both Syria and Iraq asserted the argument was political in nature. According to this professor, there would not be any hope of solution until all three parties were able to address the problem from the same foundation.

The speaker went on to give us his opinion of where Syria and Iraq had gone wrong, though again, in the absence of representatives from either of those countries, this was difficult to corroborate. He alleged that through display of misleading policies and the rejection of Turkey's proposal for a "three-staged plan" which he claimed was based on firm legal and scientific fact, Syria and Iraq were prolonging and escalating the dispute. Because Syria and Iraq continued to make this a political as opposed to a legal issue, he repeated, and what's more were working to make it a political sore point throughout the Islamic world, the parties were getting further and further apart.

If Syria and Iraq genuinely desired to settle this dispute, he charged, they should provide all the relevant technical data "on an objective and realistic basis" for the realization of an equitable, reasonable solution that would result in the optimal utilization of the waters of the two rivers. The professor pointed out that it was a well-known principle of law that it is very difficult to resolve a dispute by diplomatic means or through international adjudication without the consent of the states involved. He noted that a unilateral application for compulsory international adjudication was unlikely at that time, as there was no agreement in force between the parties calling for a compulsory settlement. The dispute, in other words, was going nowhere.

At this point in his presentation, the depth and importance of the professor's argument began to shine through his lawyerly assertiveness. He was not only trying to put into relief the complexities of conflicts relating to water; he was also giving us a real taste of what it was like to be in the midst of a bitter, ongoing transboundary dispute that threatened the geopolitical balance in one of the most explosive regions in the world. This was not a static legal case put before us like an open casket. This was a real, live, ugly dispute and we were right in the middle of it. The Turks wanted it to be crystal clear that this was no mere academic exercise in which this professor and others in the room were involved. Countries were fighting for their lives here, and for their futures. In such struggles the rules of polite protocol went out the window. If you wanted to understand conflict in transboundary water policy, just watch. This, the Turks were saying, is what their situation was *really* like.

In reflecting on this it struck me that the professor's presentation might not change the fundamental facts. Even he had acknowledged there wasn't enough water in the Tigris and Euphrates to meet each country's future needs. All the statistics and legal precedents he presented aside, economic development of the overall region would not likely be optimized if Turkey continued to think only of itself. Clinging stubbornly to national interest will only get you so far. Conflict is dangerous and unproductive. In the end, the only enduring way to be successful is to work together. Sooner or later, it occurred to me, Turkey would have to examine the value of transcending strictly legal means and interpretations in order to work out an equitable water sharing agreement with Syria and Iraq such as Süleyman Demirel had described. If they didn't, then the region might very well explode one day.

6. LESSONS FROM THE ROSENBERG FORUM

Dark and silent late last night
I think I might have heard the highway call
Geese in flight and dogs that bite
Signs that might be omens say I'm goin', goin'
Gone to Gaziantep in my mind

— James Taylor, "Carolina in My Mind"
(post-Turkey variation)

Many lessons about managing water and similar resources have surfaced, and I am sure will continue to surface, for me as a result of being involved in the Rosenberg International Forum on Water Policy. Some of these lessons have the power to alter the way I think about myself and my work. Just as my first visit to this region of the world had blown the doors of my perception wide open, so had this journey expanded the possibilities of working effectively during the next period of my life toward a sustainable future for all.

1. We have yet to define the terms of true sustainability

The problems the world is facing are growing at an exponential rate. We are arriving at solutions at a linear rate. Making the leap from environmentalism to true sustainability is the greatest intellectual challenge presently facing humanity.

2. Acting unilaterally doesn't work in the long term

Over the long term, self-interested unilateral actions create more grief than good. Strategies and frameworks not agreed upon by all partners are unlikely to produce win–win results over the long term simply because unilateral actions perpetuate disparity which in turn promotes dispute which results in conflicts that lead to lose–lose situations for everyone involved.

3. Co-operation is not optional

Attitudes and habits are hard to change. Where co-operation does not exist, progress, whether economic, social or environmental, is difficult. Unless you can afford to waste time, money and energy cancelling out one another's efforts, co-operation is a must.

4. We need to change our most basic institutions

Established precedents, including laws and treaties, are increasingly inadequate to the circumstances we are facing. To adapt to change, we need new institutional frameworks. Such frameworks are most often products of co-operation.

5. We must build trust if we want to build sustainability

It takes time to permit co-operative processes to produce the kind of trust that generates results. Trust doesn't happen overnight. Successful co-operation often requires a foundation that takes years to build.

6. It Is important to build trust before you build anything else

It is more difficult to build trust with others if you come forward with a *fait accompli*. Co-operation has to precede framework development. It is important to put participatory processes in place before projects are already determined.

7. Strong political will and support are vital

Successful prolonged co-operation is not possible without strong political will working in full support of the common good. Institutional change is not possible without political dialogue and support.

8. Broad dialogue is necessary for trust and effective action

Setting up a dialogue is necessary for bringing people together. It isn't just high-ranking business people and politicians who have to be engaged in this dialogue. The widest possible range of bona fide stakeholders has to be engaged in an ongoing way.

9. Tested solutions exist

A great deal of knowledge about past and present water management practices exists, and many of the problems are well understood. International example provides huge opportunities for the adoption of successful strategies in local circumstances.

10. We have to act now

To be successful in addressing serious interlocking environmentally related economic decline, we have to act now. To stay ahead of these problems we have to overhaul our institutions locally, regionally and nationally. It is important to build on co-operation that has already yielded successful results.

Presently, in Canada at least, there is slack in many of our systems, which means we have options. But we have to act now if we are going to stay ahead of changes that will limit our future social, economic and environment potential.

SEVEN

RIVERS OF MEMORY

"… the generosity of this planet unimaginable."

— Craig Childs, *Atlas of a Lost World*

As I will note again later in the book, what the loss of hydro-climatic stability tells us is that true sustainability may be beyond our grasp if we don't do the right thing now. Many believe that, if only out of sheer necessity, we will adapt and become more resilient as a society. But we can, and need to, do more. We keep talking about adaptation in the service of resilience; but resilience implies protecting what we have now. We need to be *presilient*; we need to protect what we have, of course, but more than that we need to adapt now for what is to come. To be sustainable, development in the future must be not only environmentally neutral also both restorative and presilient. It appears that J.B. MacKinnon agrees. His 2013 book *The Once and Future World* adds considerably to the case for restoration.

MacKinnon posits that as a society we are making a funda-mental error in our perception of the natural world, a mistake he calls "the error of the historical present." We wrongly assume that the nature we know in our time is the same nature that has always been, at least by any measure of time meaningful to our lifespan. We forget that nature was here long before us and will be here long after we are gone. Each generation, MacKinnon observes, gives itself varying degrees of permission to degrade nature as they have inherited it and know it in their time, and must live with the consequences. We are, MacKinnon argues, always just a single generation away from establishing a new sense and standard of what is normal. Yet we are largely blind to the fact that we do this.

MacKinnon is of the view that we suffer from "change blindness," caused in large measure by what our society directs its attention toward. On top of that, we have "change blindness blindness": we fail to see or even notice what we fail to see. Denial, MacKinnon argues, is the last line of defence against memory. It allows us to forget what we would rather not remember; then to forget that we have forgotten it; and then resist any temptation or inducement to remember.

In the Earth sciences, we deal with denial every day. Part of the work we do is to help society deal with uncomfortable truths. But we didn't expect to have to do this work to quite the extent that is now proving necessary. If what is happening today weren't so tragic, it might be funny. To paraphrase a former director of the U.S. National Security Agency, James Clapper, it does not matter that they know they are riding a dead horse; deniers refuse to dismount.

So what do you do when you discover you are beating a dead horse? Their strategy is to go out and buy a bigger whip. Or they change riders and then proclaim, "This is the way we have always ridden this horse."

To cover up the fact that the horse is dead, they often provide additional funding and training to increase the dead horse's performance. Or they appoint new committees of public relations and industry experts to further study the horse. Upon further study their strategy inevitably is to appoint new teams of public relations experts to revive the dead horse.

When that fails they hire outside contractors to ride the dead horse for them. When that doesn't work they lower the standards so that more dead horses can be included in their argument. Or they harness several dead horses together – to increase speed – attempting to mount multiple dead horses in hopes that one of them will spring to life.

When that fails they declare that since a dead horse doesn't have to be fed, it's less costly, carries lower overhead and therefore contributes more to the mission of sowing doubt about climate change than live horses would.

Then, finally, there is my favourite strategy: when all else fails they form a partnership between a veterinarian and a taxidermist to assure that, either way, they will get their horse back. As we have seen in the United States, as long as they stay on the horse, deniers continue to have a significant damaging effect on any influence science can have on public policy. To once again paraphrase General Clapper, national intelligence and climate change both face the same challenges.

Water and climate researchers, by virtue of their commitment to the scientific method, continue to be expected to provide integrated, timely, accurate, anticipatory and relevant information to policy-makers. But they are required to do so in such a manner that there is no risk and no embarrassment if what the research demonstrates is publicly revealed. There must also be no threat to the bottom line of any sector or individual business, and not even a hint of political jeopardy to any elected official or political party anywhere in anything scientists report. This post-truth approach to science might well be dubbed "immaculate collection."

As a society we find denial useful. It fulfills our need to at least appear to be dealing plausibly with troubling realizations. MacKinnon offers that the denial we confront today with climate change is hardly a new phenomenon. Even when now-iconic species like the dodo went extinct in the late 17th century, any assertion that it ever even existed was publicly disputed and rejected in elite circles. Similarly, when the great auk became extinct in 1844, there were people of standing who were prepared to state that "in all probability, the so-called great auk of history was a mythical creature invented by unlettered sailors and fisherfolke." There is little difference between assertions such as these and statements made by the Trump administration in the United States regarding the overwhelming scientific evidence relating to the growing human influence on the global climate. If you don't want to believe in climate change, just deny its existence.

MacKinnon offers that the "adapt and forget" pattern of change blindness is amplified by modern life. If you live in a major city, for example, you may have change blindness with respect to alterations in the natural order but nevertheless may not be suffering from environmental amnesia, because you are not likely to have memories grounded in nature in the first place. Subsequent generations, MacKinnon notes, accept as normal certain conditions their parents had to work hard to adapt to, and each generation is likely to carry forward little memory of the city as even their parents knew it, never mind their grandparents. Each subsequent generation needs new and ever more complicated mechanisms for adapting to change: more and more complex technologies and ever increasing degrees of social organization, only to respond to environmental challenges that are of our own making in the first place. Ronald Wright, in his book *A Short History of Progress*, calls this societal dilemma a "progress trap."

Part of our blindness is our obsessive focus on economic values. MacKinnon cites a situation in Maoxian County in central China in the 1990s where overharvesting by honey-hunters, excessive use of pesticides and too much land-clearing had decimated wild bee populations to such an extent that there were not enough bees to pollinate local apple orchards. By 1997 most of the growers were hiring local peasants to pollinate their trees by hand. It was, as MacKinnon notes, an object lesson in the importance of maintaining the diversity of natural species, and of how desperate circumstances can become when natural systems collapse. Not so, said three American economists who studied the response to the pollination crisis 15 years later. MacKinnon reports that the three economists published an analysis under the title "The Parable of the Bees," in which they held that the apple growers' use of hand pollination was actually more effective and efficient than relying on wild bees. In addition, the economists added, the wages paid to the hand pollinators were spent locally, further bolstering the economy of Maoxian County. The economists concluded that "destroying and replacing the free gifts of nature can be an economic benefit." So much for those economists. Well, not quite. Economists working on the fringe of their so-called science who have not lost their minds do argue that market valuation is an exercise for people who have no sense at all of how our global economy is embedded within natural Earth System function. Still, such valuations remain the norm.[3]

There is a dark lesson in this parable. People can and do survive in a degraded world. In certain instances people actually appear to thrive in degraded conditions if their economic situation can be altered to accommodate them or insulate them from change, and if environmental externalities are ignored – which is to say the costs unaccounted for by the price system of valuation, which are what we pass on to future generations by ignoring environmental degradation.

Here MacKinnon offers an interesting observation: that intergenerational memories are often stripped-down icons only. The example he gives is helpful. He notes that the Holocaust memory in our time replaces that of every ethnic cleansing before or since, of whatever scale. What are our stripped-down icons with respect to the diminishment of nature? Rachel Carson's *Silent Spring*; the *Exxon Valdez* and *Deepwater Horizon* oil spills; Hurricane Katrina; the Canmore–Calgary flood; the fire at Fort McMurray? What we are doing to nature on a planetary scale cannot be exemplified by any of these disasters. It is their collective effect that is the real story – and that is precisely the story we don't want to hear, and to the implications of which we remain wilfully blind. What I gained from MacKinnon's book that I didn't want to know but now have to face was this:

3   While I was initially outraged by this assertion, I had to admit, after thinking about it, that there could be a kernel of truth to be seen in this claim. If you have ever watched bees pollinate a fruit tree, you will have noticed it is not a straightforward process, especially in windy weather. Bees hover like helicopters trying to land in gusty conditions until there is a lull in the wind, and then, like the skilled helicopter pilots they are, they set down into the flowers. This noted, the idea that there is economic benefit in using human intervention to replace services provided by nature for free is just plain ridiculous.

nature is not the temple I have always held it to be. It is a ruin. A beautiful ruin, to be sure, but a ruin just the same.

Nature as we know it today is a fraction of what it was in the past. Despite their claims otherwise, even Aboriginal peoples are not innocent of contributing to the ruin. Researchers like Charles Kay are of the view that nature in North America before Columbus was a landscape in which Indigenous peoples had hunted wildlife into scarcity and extinction, a condition Kay refers to as Aboriginal overkill. Kay contends that Indigenous peoples were already beginning to see this for themselves, but then 1492 came along. Kay and others go so far as to suggest that the huge herds of bison found on the Great Plains after European Contact came into existence because diseases like smallpox had so decimated their main predator that populations exploded. While there is much controversy around such claims, there is no question that 1492 was the beginning of a clash of shifting baselines. While European nations were ecstatic with the discovery of apparently limitless natural riches, the continent's original peoples were formulating new, far broader understandings of ecological limits. While Kay's assessments and those of others remain contested, it is clear that even if Indigenous peoples were responsible for ecological overkill, they certainly came by it honestly. By one account, MacKinnon offers, the general attitude about nature in the Middle Ages in Europe was one of "widespread lack of generosity towards wild birds and small mammals." No kidding. Today we think the Chinese eat everything that moves or flies around them. Evidence suggests that Europeans were certainly players in that league also. MacKinnon reports, for example, that for the enthronement banquet for the archbishop of York in 1466, records show that the assembled guests were served some 400 swans, 2,000 geese, 300 mallards and teals, 204 cranes, 204 bitterns, 400 herons, 400 plover, 2,400 ruffed grouse, 400 woodcocks, 100 curlews, 4,000 pigeons, 104 peacocks, 200 pheasants, 500 partridges, 1,200 quail and 1,000 egrets. It must have been a hell of a dinner party. One wonders what abundance there must have been to permit such an extraordinary slaughter.

MacKinnon goes on to describe the bounty that was the natural world of barely 400 years ago, with the story of a French traveller named Acarete du Biscay, who visited Argentina in the late 1650s. Du Biscay, in describing what is now the megacity of Buenos Aires, noted that the river was full of fish and whales the locals called *gibars*, which today are known as fin or humpback whales. Otters too were abundant. The adjacent woods, du Biscay reported, were full of wild boars and a great many stags. He noted a remarkable story he claims was told to him by locals that underscores the great plenty that existed in nature at that time in South America. The people of Buenos Aires explained to du Biscay that from time to time their settlement – which was not yet a city – was threatened by pirates or foreign armadas. When this happened the men of the community would mount their horses and drive all the wildlife they could find in the coastal grasslands in a stampede down to the shore where the intruders were threatening to land. As MacKinnon observes, it must have been quite a scene. Imagine a thunderous stampede of tortoises (an apocalyptic herd of turtles?), snakes, lizards, voles, mice, armadillos, foxes, wildcats, ground birds, songbirds and locusts, all heading for the shoreline. The stampede, du Biscay was told, was such a frightening thing to behold – a veritable impenetrable wall of wailing, barking and buzzing on the shoreline – that it drove buccaneers and even foreign armadas back out to sea. Could anything like that happen now? MacKinnon puts it this way: Can we even imagine wildlife so plentiful it could be used as a military defence?

I know the natural abundance of where I live was already dramatically diminished by the time I came into it. For years I studied the fur trade, but largely from the point of view of exploration, mapmaking and contact between peoples, not in terms of its impacts on the natural systems of what is now called the Canadian West. But even if you stick with the fur trade as a study in geography, the fact of its effects is impossible to ignore.

Imagine a nation being created because of a fashion trend on another continent. We live in a country that came into existence largely because of a rat and a hat. Canada was "ratified" and "hatified" as a nation by the fur trade. It is hard to explain to others that before we were a nation, we were essentially a Hudson's Bay department store. The whole world "shopped at the Bay."

The main reason for the fur trade in Canada was the increasing demand for beaver hats. Historians have estimated that by the 17th century the market for hats in England alone was nearly five million units a year. Over the next century, population growth and the expansion of an export market in Europe resulted in rising demand for beaver pelts. It is estimated that over the 70 years to 1770, 21 million beaver and felt hats were exported from England. The impact on Canadian beaver populations was catastrophic. It is estimated that from York Factory alone between 1716 and 1770, an average of 35,000 pelts were being shipped to England every year. In big years like 1730 and 1731, the average increased to more than 55,000 pelts. When a given region became depleted of beavers the trade simply moved on.

The fur trade affected wildlife populations far beyond what was harvested for furs. Jasper House, an early fur trade post located in what is now Jasper National Park, for example, took a profound toll on wildlife of almost every kind in the Athabasca Valley. Prolonged hunting pressure from supplying food for Jasper House traders and their families and passing brigades began over time to seriously reduce bighorn sheep and moose populations. As I noted in *Ecology & Wonder,* Father De Smet's account of his visit to Jasper House in 1846 gives us an idea of the nature of the problem. He observed that in 26 days, Jasper House hunters killed 12 moose, 2 caribou, 30 bighorn sheep, 2 porcupines, 210 hares, a beaver, 2 muskrats, 24 geese, 115 ducks, 21 pheasants, a snipe, an eagle, an owl, 30 to 50 whitefish a day and 20 trout. It is not surprising that a few years later the game was gone and the Hudson's Bay Company had to prohibit freemen from hunting within 50 kilometres of Jasper House. How long did it take for wildlife populations to recover from the nearby presence of a fur trade post? We don't know if they ever did.

MacKinnon cites the observations of the famous Scottish botanist David Douglas, who was witness to the slaughter and the mindset that attended it. Douglas, outraged by a landscape empty of any fur-bearing animal larger than a chipmunk, told the chief factor at a Hudson's Bay post he visited that "there was not an officer in it with a soul above a beaver skin." Though "Bay Days" are over – bought up, along with Canadian Pacific hotels, by foreigners – the beaver remains the symbol of our nation. And what a symbol it is.

MacKinnon has observed that we lose not only words but whole phrases if we are too busy to observe what they mean, or when what they represent disappears, as a consequence of intergenerational neglect leading to loss of memory or blindness to change. MacKinnon puts forward an entire class of phrases under various levels of threat, including expressions such as school of fish; pride of lions; tiding of magpies; kindle of cats; and my favourite, an unkindness of ravens. And how about this one? A shrewdness of apes. Such expressions are not just quaint; they speak of a different, more intimate way of relating to these creatures based on seeing them not just as animals but as other beings with their own unique character and behaviour. Such expressions are born not just of identification but of observation over long periods of time of generations of other creatures with whom we share our existence. Relations between us and the other life forms these expressions signify are breaking down in a diminished natural world.

As much as we might like to think otherwise, we are all of us complicit in this diminishment. We are all in this together. While it remains open to debate whether Indigenous cultures in North America had the numbers, knowledge and power to transform an entire continent, a widely treasured and deeply held ideal of true wilderness is now being challenged. Nature as it has been defined – which is to say the sum total of everything that is not us and did not spring from our imaginations – is under siege. Nature is now seen to include both us and everything that does in fact spring from our minds. It also includes everything we do that affects Earth System function. MacKinnon is quite right. There is no question there was more of everything natural in the past because there were fewer of us. The cascading effect of losses over time appeared to us at least to be gradual – almost invisible – and they remain mostly unknown and unknowable today. But those losses were and are substantial. They include extraordinary damage to coral reefs that resulted in the loss of species and reduced numbers. This diminishment has had an even more profound effect on ocean fisheries, both in terms of populations and of the size of individuals of each species. Almost all fish caught commercially today are smaller than they were in the past, with a complete decline in the numbers of larger individuals. Generations in the past would be shocked at how small commercial fish catches have become and at the small size of the individuals found in today's nets. To them it would appear we are netting juvenile fish that have not been allowed to mature to adult size. This, as MacKinnon notes, is easy to track. All you have to do is look at photographs of catch size in both commercial and recreational fishing over the past half century.

I have seen this personally with the size and vulnerability of west coast oysters. A year ago I agreed to speak at a conference in Comox, on Vancouver Island, with the proviso that at some point while I was there I would have an opportunity to enjoy the region's famous Fanny Bay oysters. Unfortunately, the commercial beds from which these oysters were harvested had been closed as a result of contamination from agricultural runoff flowing into the waters of the bay. But the Comox First Nation saved the day. Upon hearing of my disappointment, people from their reserve graciously picked me up at my hotel and drove me at low tide to the natural oyster beds in their territory, which had not been contaminated. Unlike the oysters grown industrially and sold while still very small, the oysters my hosts shared with me were huge. Some were almost as large as a dinner plate.

This is completely consistent with the observations MacKinnon reports. He notes that some shellfish have steadily declined in size over the past 10,000 years of human predation.

Red abalone have gone from the size of the oysters the Comox people shared with me to the size that would permit a dozen to be served on one plate. Tiger sharks off the east coast have gone from a typical eight feet in length to half that size, as have many of the most popular trophy fish. It is estimated that 80 per cent of the world's reefs had their large animal species depleted before 1900. The number of coral reefs globally that are now considered "pristine" is zero. Some 30 per cent of the world's coral reefs are within marine protected areas, but only 6 per cent are thought to be effectively protected from pollution and overfishing. And it is not just the reefs that have been decimated. The world's largest fish, from tuna to cod to swordfish and sharks, have been reduced to 10 per cent of their former abundance. The great whale populations have been similarly diminished. The diadromous fish – those species that divide their lives between fresh and salt water, including salmon, herring, eels, whitefish and sturgeon – are among the hardest-hit species on Earth. Nor are fish the only ones in trouble: so are the waters they swim in. Current estimates are that 40 per cent of our rivers are in a deteriorated state, and that a third of freshwater fish species are extinct or threatened.

Every corner of the planet has been affected by human influence. While 14 per cent of the terrestrial Earth is currently under some form of protection, most of this area is comprised of desolate landscapes or covered by snow and ice. Only 5 per cent of temperate grasslands are protected in any meaningful way. We are down to a 10 per cent world. But biodiversity loss is only part of the story, which must now include climate change, land-use impacts and global soil deterioration and loss. The cascading effects are suddenly becoming knowable and terrifyingly obvious. We find ourselves in the throes of a sixth global extinction, which we ourselves have brought about.

There is no question we are in the midst of an accelerating ecological free-fall. Humanity is now a cataclysmic force in its own right, right up there, as MacKinnon says, with the killer asteroids and glaciers three kilometres deep. It is not just the loss of individual species that is of concern; it is the loss of biodiversity itself. As of 2013, 31 per cent of all amphibians, one in five mammals and 13 per cent of all birds were on the brink of extinction. Not much has changed since. This is not a trifling matter. MacKinnon reminds us that extinction wipes out, point by point, the clues to Earthly existence, and that extirpation "is the great sucking retreat of the tide of life." We now know there is a direct link between extinctions and climate change. Climate change in our current context is a consequence of damage we have done to other parts of the Earth System, among them our impact on biodiversity. Climate is very much influenced by biodiversity, in large measure because of the combined effects living things have on the atmosphere. There is evidence in the fossil record of a rise in extinctions in our time. Extinctions are presently occurring at a rate a hundred to a thousand times faster than in background strata. Well, you might say, doesn't every species eventually become extinct? Yes, but it is the *rate* of extinction, just as it is the rate of climate change, that matters. The reality is that with extinctions in the past, species often didn't die at all, but instead evolved into two or more sister or daughter species. The accelerated extinction rates brought about by human impacts do not allow time for new species to emerge from the old.

The problem here is that our planet's self-regulating Earth

System – the system we depend on to provide the predictable conditions our economic and social stability require – is biodiversity-based. If you take out too many parts, its function will change or cease. At the moment, we are not just diminishing the stability of the Earth System; we are dismantling it. The most vulnerable habitats of all, with the highest extinction rates per unit of area, include rivers, streams and lakes in both tropical and temperate regions. MacKinnon shares these concerns. He puts it this way: "Extinction is not mere death; it is the death of the cycle of life and death."

As sociobiologist E.O. Wilson has observed, we are on our way to being alone in the world. These threats suggest there has perhaps never been a more important time to recognize and act on the fact that ecosystems do not exist simply to create benefits for one species and one species alone: humans. Given we barely know half the organisms that comprise our ecosystems, sustaining ecological benefits will require precautionary care if only because we don't know how to replace many ecosystem services with technology, even if we could somehow afford to.

When we speak of sustainability, we don't mean just the present. We are also talking about the past. The natural world of antiquity is not simply lost and forgotten. The matter is more complicated than that. If you are unaware of past presences in nature, their absence might appear perfectly natural. But absence can also have a presence. The idea of the presence of absence dates back to Plato. It is a concept that challenges us to undertake an inventory of the missing. As MacKinnon points out, evidence of absence is all around us. Why, he asks, is the pit of an avocado so disproportionately large? It is that size because its evolutionary purpose was to pass intact through the digestive tracts of creatures larger than any animal living today. I myself have told some of the same ghost stories MacKinnon shares.

While most of us would hardly consider species disappearances relevant in the context of our time, the ghosts of missing predators are with us today. Consider the pronghorn antelope. If you have seen one in the wild, you may have wondered why the pronghorn is capable of such extraordinary bursts of speed – far faster than any predator hunting the pronghorn today can achieve. As it happens, there was a time when the pronghorn was the central prey species of the North American cheetah, which became extinct at the end of the last ice age. If you live in the West you may have seen pronghorns trying to keep up with passing cars. One wonders if they miss the Pleistocene cheetah that once chased them hungrily across these same plains. Or if they are simply being pursued by ghosts.

It appears also that two bear species disappeared in the Pleistocene that were much larger than the related species that survived them and that form the wildlife assemblage we know today in western North America. The giant short-faced bear, *Arctodus simus*, was probably nearly 4 metres (13 feet) tall when it stood up on its hind legs. It was larger than any bear that survives today, including the polar bear. This bear would have attracted considerable attention when it suddenly stood up among willows by a river. When early North Americans confronted this bear – as surely they must have – they would have found themselves in the presence of one of the great natural symbols of the Pleistocene and beheld the spirit of the greatest of the Great Bears. Through knowledge of this animal they would have been able to ceremonially define just how much wild there was in the wilderness they claimed as their home.

Though smaller, today's grizzly bear continues to perform this age-old role for those who live in the West today.

The continued presence of the Great Bear enlivens us and enriches our experience of place. If for some reason the grizzly were to disappear, we would be like the pronghorn in the absence of the cheetah. We would wander our Western mountains missing that element of landscape that demanded the most of us: the most courage, the most cunning and the most understanding of who we are and of the powers of the ecosystems we depend on not just for our survival but for our identity. We too would be pursued by ghosts.

You have to pay attention if you want to observe the presence of the absent in our contemporary world. With the exception perhaps of some Indigenous peoples, we as a society and as individuals are so increasingly distracted from nature that we do not see the loss of a flower, a tree, a fish, a bird or any other animal as a story we can no longer tell. But we are beginning to understand just how complex extinction really is on a broader scale. Species don't just go quietly into extinction and that is the end of it. There are deeper impacts. When a species disappears we don't just lose an Earthly companion which we may or may not have interacted with or even known. We also experience knowledge extinction. When landscapes are diminished and species are lost we are deprived not only of knowledge but of Earth System capacity and resilience. A good example is our growing vulnerability to climate change. We are not facing the serious consequences of climate disruption just because human-generated carbon dioxide concentrations are rising. Though that is certainly part of it, it is not nearly the whole story. We are also in trouble because at the same time, as MacKinnon points out, we have stripped down a transformed nature to such an extent that it can sequester only a fraction of the carbon dioxide it was able to handle in the past.

It is this degree of impact that led Paul Crutzen and Eugene Stoermer to declare that we have now created and entered an utterly new, human-induced geological epoch, defined for the first time in the history of this planet by the fact that humanity is having a greater effect on the functioning of the Earth System than nature itself is. MacKinnon outlines the extent of some of these impacts. I have updated his list.

Among the factors that define the Anthropocene epoch are our numbers and the numbers of animals we need to feed ourselves and sustain our way of life. At the time of this writing (12:20 p.m. Eastern Standard Time on Thursday, June 7, 2018) there were 7,627,492,069 people on Earth and counting. According to statistics from the UN's Food and Agriculture Organization there were at the same time almost 19 billion chickens, or about 2.5 per person. Cattle are the next-most-populous breed of farm animal, at 1.4 billion, with sheep and pigs not far behind, at around 1 billion.

More than half of the surface of the terrestrial Earth has been altered by human activities and agriculture. As a society we pump more sulphur dioxide into the atmosphere through the burning of coal and oil than all the natural sources of that gas combined. We have created a permanent new atmospheric and weather condition called smog. We have polluted the planet's atmosphere with carbon dioxide emissions to such an extent that even if we stopped these emissions at midnight tonight, we have likely already affected the Earth's climate for at least the next 50,000 years. MacKinnon notes that even if we put down our tools today, we will still end up with a world of our own creation. As we plunge pell-mell into the Anthropocene the choices we make are going to have to be our own. No matter how uneasy some of us are about our own hubris or how frightened or unwilling we might be to shoulder responsibility for the rest of creation, it is we who have trashed the planet. And if we don't want to live in degraded circumstances of our own making for the rest of time, we have no choice but to clean up our mess and rehabilitate the world.

The way you see the natural world around you determines much about the kind of world you are willing to live with. The nature we want to live in is a choice. But unless we start paying attention, our choices will be increasingly limited and ever more unaffordable. Our own ancestors handed down an already vastly degraded world. We have in this generation accelerated that degradation. We are now faced with a critical question: Where, in our time, do we draw the line?

We have already proven we can live in a much diminished world. We have shown we can probably adapt to whatever reduced conditions we ultimately create. But is that what we want? As environmental historian Donald Worster has written, there has probably never been a time in the life of our species when holding fresh the memory of the world has been more urgent. For it is only against a memory of the world that we can measure the real progress of our civilization and the validity and desirability of its direction. This is not a new idea. An 1864 book called *Man and Nature*, by New England scholar George Perkins Marsh, called for the restoration of the world. What, one wonders, would the world look like if we restored it to some semblance of its earlier natural abundance?

MacKinnon reports that Tony Pitcher of the University of British Columbia has thought about what a restored global ocean could do for humanity. His back-of-an-envelope estimate is that if we could somehow miraculously rebuild fish stocks to their historical bounty, we could catch 10 per cent of the fish every year without a decline in abundance. That amount of fish, Pitcher projects, would be equal to 40 to 60 per cent of the current annual catch worldwide today. Can you imagine? If we could get back what we've lost we could have a sustainable global fishery. Instead of losing 90 per cent of our bounty, that same bounty would sustain us. Instead, we live in a 10 per cent world.

MacKinnon also offers interesting perspectives on the role whales play in carbon sequestration. When whales die, their bodies sink to the deepest abysses of the ocean, taking with them all the carbon they contain to depths where full decomposition could take centuries. If we restored the great whales to their former populations, how much carbon could we sequester and still have the presence of whales to inspire us? Researchers have demonstrated that even today's decimated population of sperm whales alone removes some 240,000 tonnes of carbon from the global atmosphere each year. If we brought sperm whales back to pre-whaling numbers, the amount of carbon they would collectively remove annually – and remember, this is just one of the 13 species of great whales that are struggling to survive in the global ocean – would reach 2.4 million tonnes, which, as MacKinnon noted in 2013, would have been valued at more than $20 per tonne on the global carbon exchange that existed at that time.

Though it is difficult to imagine such restoration occurring on a global scale in time to avoid runaway climate change and irreversible Earth System damage, MacKinnon for one holds the optimistic view that restoration of parts of Earth System function could even be possible with a global human population of seven billion. This would require a massive program of rewilding, however. And of course it immediately raises other questions. Could we live in a rewilded world? Would we want to? Restoration would demand the reintroduction of many species that have largely been exterminated or extirpated because they either threatened us or competed with us for resources. On this matter MacKinnon offers an interesting observation: it is not fear that drives us to extinguish fearsome creatures, but once they are gone it may well be fear that keeps us from bringing them back.

Having made the case for protection and restoration as critical elements in our collective efforts to sustain the conditions that make human existence not only possible but meaningful on this planet, I am grateful for J.B. MacKinnon's thoughtful analysis, example and company. In reading his book, it struck me that we need to do more to identify, connect and advance what I call the interdisciplinary restoration sciences – restoration hydrology, aquatic ecosystem restoration, and rewilding in the context of conservation biology.

The connecting of these emerging disciplines has to occur quickly and with full recognition that if we fail to restore the world, the Anthropocene is going to be a humbling if not dangerous experience for all of humanity. MacKinnon reminds us that paying attention is the foundation of restoration. He also reminds us that while nature may not be what it was, it hasn't simply gone away, either. As MacKinnon discovered while swimming in the estuary of the Guadalquivir River in southwest Spain, nature may be diminished but it is still there – waiting. Beyond this critical reminder, MacKinnon poses some of the most important questions of our time and then answers them.

We have continued in our time the degradations of the past. We are faced now with a difficult question: Where do we draw the line? But even when we agree we must restore what we have damaged, we find ourselves debating what goal or era or condition our effort at restoration should aim for. Forget the baseline for the moment. It is not the baseline that matters right now – it is the direction.

We need to step back from the Anthropocene abyss. We need to buy the time we need to achieve sustainability. Only by restoring the natural world to some semblance of its former abundance and diversity can we buy that time.

We need to repudiate denial and overcome our intra- and intergenerational blindness to change. For it is only against a memory of the world and its past abundance that we can measure the real progress of our society and judge the validity and desirability of its future direction.

EIGHT

RIVERS OF ICE

I have arrived in Santiago, Chile. I am here at the invitation of the University of Saskatchewan's Dr. John Pomeroy to attend the annual workshop of INARCH, the International Network for Alpine Research Catchment Hydrology. Having been to the three previous iterations of this remarkable scientific forum, at Canmore, Alberta; Grenoble, France; and Zugspitze, Germany, I am now headed for the fourth, at a ski resort at Portillo, high in the mountains above Santiago.

Though Santiago has some rough neighbourhoods, as all cities do, this is arguably the safest large city in all of South America. Though urban hives have lost their appeal for me, I at least feel comfortable here. In defence of Santiago, I would characterize it as more European than North American, but it discloses its South American nature very quickly. I arrived early on a Sunday morning and was very efficiently transported to my hotel in the Providencia district at speeds of up to 140 kilometres an hour. We literally blew past the horrific slums that border the heavily engineered river that runs through the city. This river – the Mapocho – has a habit of flooding, hence its confinement between concrete banks. My hotel was close to the city's most famous landmark, the Costanera Center Torre 2, better known as the Gran Torre Santiago, or Great Santiago Tower. This 64-storey skyscraper is the tallest in all of Latin

America. It is interesting, however, that a swollen Mapocho River flooded the underground shopping centre beneath the Gràn Torre, delaying completion of the building for more than two years. You have to love wild rivers.

I had a day to myself in Santiago before Dr. Pomeroy and his colleagues Chris DeBeer and Sean Carey arrived from Canada and Ethan Gutmann and Tom Painter from the u.s. When John arrived he was under the weather, and the next day he was quite ill. I stayed with John the next day while the others met at the University of Santiago and then travelled two hours by motorcoach, winding slowly up the steep mountain pass that formed Chile's border with Argentina, to the conference venue at the ski area of Portillo. John and I were picked up the following morning by renowned Argentine glaciologist Dr. Gino Casassa, who was also attending the conference.

While all the presentations were valuable, two were of particular interest in the Canadian context. These concerned the monitoring of the state of glaciers in Chile and Argentina.

My recent specific interest in the diminishment and loss of glaciers in the Andes was much heightened by reading a book I bought, oddly enough, in the middle of the Canadian prairies, at Tergesen's legendary general store in Gimli, Manitoba. In *Glaciers: The Politics of Ice*, Jorge Daniel Taillant outlines the events and politics associated with Argentina's crafting and passing of the world's first glacier protection legislation, in 2010.

As Taillant tells it, the story began with a meeting in the office of Argentina's federal secretary of environment, Romina Picolotti, on September 6, 2006. The meeting was called to hear environmental concerns expressed by locals affected by the development of the world's first large-scale gold-mining project straddling an international boundary, in this case the border between Argentina and Chile. One of the major problems with the project was that its proponent, the Canadian mining giant Barrick Gold, required removal of an estimated 10 hectares (25 acres) of glacial ice in the upper reaches of the Rio Toro basin, ostensibly to avoid slope instability and other environmental impacts during the life of the Pascua Lama open-pit mine. As it happened, our conference was being held just below the same Continental Divide separating not just Chile and Argentina but the waters flowing to the Pacific Ocean on the Chilean side of the Andes from those flowing ultimately to the Atlantic by way of Argentina. (And what a stunning divide it is.)

At the time of the meeting, Picolotti, whom I once had the pleasure of meeting, was in the midst of revamping Argentina's Environmental Secretariat in the wake of a case she had helped Argentina put forward to the International Court of Justice at The Hague against neighbouring Uruguay. The case concerned a unilateral decision by Uruguay to construct two major pulp mills, financed by the World Bank, on the Uruguay River, which marks the boundary between the two countries. Picolotti's level-headed, strategic approach, and the success of the case, had so impressed the president at the time, Néstor Kirchner, that he appointed her as environment secretary.

Picolotti's meeting on September 6, 2006, was scheduled with a group of some 15 environmentalists from San Juan province, in a remote part of Chile Taillant describes as resembling the American Far West, where the huge Pascua Lama mine was located. The group had swollen to nearly 50 by the time they arrived by bus in Buenos Aires. They were concerned about the environmental damage of such a large mine. They were also troubled that there was little or no public control over these impacts. Their biggest concern, however, was the potential effect of the mine on the region's precious water supplies. The slogan of the group – which later echoed across all of Latin America – was "Water is worth more than gold."

Still, there probably would not have been any issue, and Taillant might not have had a reason to write his book, had it not been for a rumour that Barrick wanted to dynamite glaciers at the Pascua Lama site, and had the company not immediately mobilized politically as it did against any expanded regulation of its mining activities. Before Barrick went political on the issue, few understood or cared about mining impacts on glaciers. Once public attention was focused on the Toronto-based miner's border-straddling operations, however, it was found that there were also glaciers and related periglacial features on the 180-kilometre dirt road to the Pascua Lama site.

When pressed on the matter, Barrick hired two glaciologists to examine effects of its operations on local glaciers. As Taillant points out, this work was meant only as a preliminary review, not a formal environmental impact assessment. According to Taillant, the report was vague and inconclusive. Environmentalists nevertheless took the report to mean that the chance of glaciers surviving in the area to be affected by the proposed mining was practically zero. Barrick countered that its impacts on glaciers didn't matter, as they were on their way out anyway as a consequence of climate change. "Not so fast," said the Environment Secretariat. "Aren't glaciers the lifeblood of the headwaters of some of the most vital rivers that have their origins in the Andes? Didn't one of the affected watercourses run through a UNESCO Biosphere Reserve, and didn't this same river supply the rich agricultural lands in the San Juan lowlands?"

Further inquiry revealed that while Barrick's study had identified only a handful, there were probably 50 glaciers in the area that could be affected one way or another by the Pascua Lama operations on the Argentine side of the border alone, plus innumerable rock glaciers. Add the area affected in Chile and the number of glaciers rose into the hundreds, with uncountable numbers of debris-covered ice masses and rock glaciers. No one could say for sure how many there were, because no inventory had ever been taken. What was known, however, was that even a very small glacier holds a substantial amount of water, and that glacial melt contributes significantly to base flow in streams after the spring melt. It was also known that glacial melt was vital to the perpetuation of high-elevation wetlands and critical to the survival of flora and fauna in the dry alpine regions of the Andes. It was held that these benefits were becoming even more important as the climate warmed, and that threats to these benefits were exacerbating a growing water-security crisis in many South American countries.

It is interesting to note that the justification for federal jurisdiction over glacier protection – as opposed to leaving it up to individual provinces to sort out – was the fact that the impacts were transboundary in nature and had international implications. There were also, it was discovered, international precedents elsewhere that called attention to how complex this issue could become. What happened at Pascua Lama, for example, was compared with what another Canadian mining company, Centerra, had done to glaciers at a site in Kyrgyzstan where sterile rock piles were dumped on an active glacier. It was also pointed out that yet another miner, Codelco, was doing something similar in glacial circumstances near Chile's capital, Santiago. Taillant asked an obvious question: Why was no other similarly affected nation doing anything about it? Here is how his book puts the issue:

We would think that, because glaciers house the most important natural resource to our existence, and because they are vulnerable and because we are unduly impacting them, as a society, we would have taken steps by now to protect them. And yet, despite the obvious vulnerability of glaciers and the fact that they are critical to the sustainability and livelihoods of dense human populations, until very recently, *not a single country anywhere in the world had enacted a national law to protect glaciers*, the veritable water towers and reserves of our planet. Glaciers, for the most part, are not protected by law or even public policy. [Emphasis in original.]

How is that possible? In an age of deteriorating planetary ecology – and with glacier retreat as one of the most visible signs of climate change – wouldn't the Canadians, the Swiss, the French, the Americans, the Chileans, or Argentines, or the Chinese, Pakistanis, or Nepalese, or the Californians who have been environmental mavericks on so many fronts and with many well-known glaciers in their environments – wouldn't someone have identified this delicate and fundamental natural resource and concluded that these resources should be protected not only for their majestic beauty, but also because they are fundamental sources of freshwater for local populations?

Taillant goes on to ask why the UN hasn't brought these concerns forward through the Framework Convention on Climate Change. Shouldn't there be a global treaty to protect glaciers? Wouldn't ski countries such as Switzerland or France, or Canada for that matter, want to put forward the idea of protection for their glaciers, if only to preserve their tourism sectors?

One of the answers to that question, in Chile and Argentina at least, is that such protections ran up against huge opposition from a very powerful mining lobby, companies with significant roots in Canada. From a legal perspective the mining lobby had plenty to work with. What water laws existed dated back more than a century and didn't so much as mention source waters, glaciers or even snow or ice. In the definitions associated with these outdated laws, ice, in Argentina at least, was not legally defined as water. The best that European laws appear to offer is that glaciers fall into the jurisdiction of governments because they are part of the public domain. In the United States and Canada the only legislative references to glaciers pertain to what national parks were created to protect. When Chile decided in August 2005 to propose a glacier protection law based on the effects of mining on or near glaciers, particularly the increased turbidity of downstream flows, the bill never made it through Congress. The mining sector killed it.

In the meantime, however, the matter had caught the attention of Alejandro Iza and Marta Brunilda Rovere of the International Union for the Conservation of Nature, who after conducting research, published a compendium of laws and regulations that could be applied to the protection of glaciers. The resulting work focused on seven countries – Argentina, Bolivia, Colombia, Chile, Ecuador, Peru and Venezuela – and concluded that none of the study countries provided an adequate legal and regulatory framework or sufficient public policy to protect glaciers. Further reports and books make the point that in many parts of the globe glaciers will be the first big loss in a warming world. Interestingly, Canadians were on both sides of the debate. On one side were mining companies like Barrick Gold and its president and founder, the late Peter Munk, who were able to lobby at the presidential level in many South American countries against what the industry considered to be unnecessary restrictions on economically important mining developments. On the other side of the issue were a German named Alexander Brenning and the Chilean Guillermo Azócar, both of whom were glacier experts at Waterloo University in Canada, who used their expertise to characterize mining impacts on glacier ice from an environmental perspective. Though they would not have been able to anticipate it at the time, their research and publications would greatly influence legal discourse, particularly in Argentina, and eventually contribute directly to the crafting of ultimately successful glacier protection legislation.

Beyond better on-the-ground monitoring of the environmental effects of mining in glaciated and periglacial regions, by far the most important means of understanding the extent, nature and significance of mining impacts was the emerging availability of high-resolution satellite images on Google Earth. While mining companies could assert whatever they wished in community engagement sessions and meetings with political leaders, the satellite images told the real story. They showed what was there on the ground. In the midst of this it appears from Taillant's account that companies like Barrick Gold continued to make themselves ever more unpopular by closing their mining roads to locals, to the point of barring mountaineers and even tourism companies from access to surrounding mountains. According to Taillant, mining companies that acted this way did not see that such tactics only fed the fire of what he described as the birth of cryoactivism in South America and elsewhere. For their part, the strategy of Argentina's Environment Secretariat was to keep the provinces on side by agreeing to pay for a national glacier inventory, the logic being that if you pay to implement a policy, you get to decide on that policy. What happened next, however, surprised everyone.

On October 22, 2008, the Argentine Congress unanimously passed the world's first glacier protection bill. Taillant explains that the reason the bill succeeded was because most of the elected representatives either hadn't read it or didn't understand its full implications. As would be expected in such circumstances, there was a brief calm before the public and the mining sector woke up to what the legislation meant. Then the shock wave hit. It suddenly became clear that few elected officials even knew what a glacier was, or even knew there were glaciers in the provinces they represented. If they had known glaciers were important, they would have paid greater attention to the bill.

When the back-pedalling began, two issues came to the fore. The first, not surprisingly, was jurisdiction. The second was the financing of the regulatory regime set out in the bill. The mining companies knew they had to move fast to catch up with what had happened and determine the best political levers to pull to ensure their interests were recognized and served. They knew the president had only ten days to veto the bill. They would have to get to the highest levels of the executive in government immediately if they were to influence the president to veto. It is at this stage in Taillant's narrative that the image of the Canadian mining sector is put into most unfortunate relief. It appears from the information he provides that the value of Canadian mining shares on the Toronto Stock Exchange is based heavily on the projected market value of reserves determined to exist when a mine ultimately goes into actual production. In other words, mining stocks essentially go up in value based largely on speculation. The object of shareholders is often to get rich on the growing value of the stock, not on the actual return from the mineral once it is actually mined. This evidently is how venture capitalism works. If the find proves big enough, exploration companies get bought out by bigger companies like Barrick Gold, which then find the most economical way to realize the profit. Sometimes these finds are actually underneath glaciers, which, until Argentina passed its glacier protection law, were considered by mining companies as just another kind of overburden that had to be removed to get at the pay dirt.

In fairness to the mining sector, it must be said that some companies do try to minimize their environmental impacts. To do so, however, they need expectations to be clearly spelled out well in advance. The sudden passage of the glacier protection law in Argentina put mining companies in an untenable position. Especially troubling was Article Six of the statute, concerning suddenly prohibited activities, which included any operations that could affect the natural conditions of a glacier or surrounding periglacial environments, including all forms of active rock glaciers. Under the new law even installing infrastructure on glaciers or in periglacial areas saturated with ice would be prohibited. Barrick Gold claimed reserves worth billions of dollars that were buried under exactly these kinds of glacial and periglacial terrains. Fortunately for the company, Mr. Munk had close ties to the president of Argentina, ties he evidently had continued to strengthen ever since the company arrived in Argentina in the 1990s to announce its investment in what was at the time touted to be the largest gold mine in the world. Barrick understood fully that if the glacier protection bill were passed, Pascua Lama might not be able to proceed.

There were evidently all manner of rumours about how Munk made his concerns known to senior levels of the Argentine government. All that is actually known, however, is that only a few days after Argentina's glacier protection bill was passed, it was vetoed by President Cristina Fernández de Kirchner. The president dealt with the political blowback of what was being trumpeted by opposition parties as the "Barrick veto" by claiming that those who voted unanimously for the bill didn't know what they were voting for. Taillant argues this was a thin excuse given that the interests that were suddenly rejecting the measure were not those who voted unanimously in favour of it and who continued to support it after it was vetoed. The reality is that the issue did not fall out exactly as glacier protection advocates would like to think it did. The estimated value of the Pascua Lama mine had risen to US$50-billion by 2011, so there was in fact a significant argument in favour of a clearer understanding of the economic, legal and ultimately political consequences of passing such legislation. The opposition uproar also put another shortcoming into clear relief. The passage of the bill showed how disconnected Argentine government departments were from one another. The federal mining secretary appeared completely out of the loop leading up to passage of the bill. Nor did provincial governments see it coming. The corporate sector also appeared out of touch with the political processes that led to the crafting and passage of the legislation. Though Taillant doesn't say it outright, it would appear that there were a number of very influential people at both levels of government as well as in the private sector who should have been highly embarrassed by the unanimous passage of a bill of this consequence.

On the other hand, it makes some sense. When meetings organized by the environment secretary resumed after the veto to determine how the measure might be revived, the mining sector was still not invited. This would be like Environment and Climate Change Canada orchestrating talks on a pending glacier protection law and not inviting Natural Resources Canada to attend, even though the Geological Survey of Canada and its Glaciology Section are housed in NRCan. It would appear that the clear message was that legislators did not want the mining sector at the table because they knew that if they were there they would neutralize the effectiveness of any legislation that would survive such deliberations.

As it happened, however, a carefully drafted industry report claiming Barrick's Pascua Lama operations would not affect glaciers or periglacial areas did throw a wrench into early efforts to revive the legislation. But the report was soon dismissed as a red herring. As the mining companies claimed the report was irrefutable, the environment secretary decided not to debate its conclusions. The view was that if the mining sector wanted to believe its own report they were welcome to do so, but the facts would ultimately prevail. And they did. It didn't take long for Barrick to figure out that simply exhibiting evidence they felt was adequate to demonstrate that their operations would be exempt from such a law wasn't going to work. If it was a bluff, it failed. That said, negotiations aimed at crafting a new glacier bill could not ignore reasonable issues that would have to be addressed constructively if the resulting revised text was to have any hope of passage and meaningful implementation. Lawmakers in Argentina faced exactly the kinds of issues that such legislation would encounter in Canada. Several interests opposed absolute prohibition of industrial activity in glacial and periglacial areas. Provincial governments opposed federal discretion and jurisdiction in the implementation of such a law, which they felt was an encroachment on their authority. That said, however, none of the provinces appeared to want to reject the federal offer to bear the budgetary costs associated with implementation should such a law be passed in some form. These issues once again put into relief an important public policy truism: getting legislation passed is one thing, but keeping it in force under subsequent governments is quite another. Getting a law enacted and getting it meaningfully implemented are also two different things. As Taillant points out, the devil is always in the details, in this case in the frozen ground of Andean periglacial regions.

Among the first details that bedevilled the process was the length of time required to do a proper glacier survey. It was deemed impossible to conduct an inventory of adequate scale and detail in only five years. In the midst of these deliberations, President Fernández de Kirchner effectively gave Romina Picolotti an ultimatum. Unhappy that her environment secretary had blindsided her with the passage of the glacier protection bill, the president made it clear through intermediaries that Picolotti could either provide supportive legal justification for the presidential veto or be relieved of her position. To her great credit Romina Picolotti decided she would not, even if it cost her political career, stand by a president who in the view of many had sold out to Barrick Gold. At 38 years of age, Picolotti graciously and honourably stepped down. And, as one would expect in problematic political circumstances where hanging on to power means everything, the president accepted Picolotti's resignation, citing as the reason for her departure the poor handling of the Environment Secretariat's budget. By this time, however, getting rid of Picolotti did not significantly change the political calculus that was beginning to crystallize around the desire for a new glacier protection law that would replace the one vetoed by the president. Picolotti had already orchestrated a huge political win. A lot of politicians who previously knew almost nothing about glaciers now had new appreciation for ice and frozen ground. Suddenly the whole country appeared to be interested in glaciers. An obscure federal institution dedicated to glacier research had suddenly come to the fore and was playing a big role in shaping public understanding of ice and snow.

The Argentine Institute of Snow Research, Glaciology and Environmental Sciences, or IANIGLA as it was known, was teaching Argentinians exactly what Canadians will need to know if Canada is to see the value in protecting its own glaciers and periglacial environments. IANIGLA showed the nation that glaciers were important because they stored water and because they regulated stream flow long after the spring melt. The science IANIGLA shared with politicians and the public demonstrated beyond a doubt that glaciers were critically important to hot and dry provinces with high-mountain headwaters, and in so doing showed it was in the national interest to protect them. IANIGLA's work established that there were glaciers hidden under rocky debris, and that other frozen environments also contained ice. The research institute also successfully demonstrated the kinds of impacts that could damage glaciers and diminish their hydrological value, and showed it would be prudent for the country to take stock of its glacial and related periglacial resources to determine their hydrological value now and into the future. IANIGLA underlined as well that much more needed to be known about glaciers and related resources so they could be protected and managed in the national interest.

But even with solid science available and good educational programming in place, there would still be serious challenges. There were misunderstandings of the intentions of the legislation and misinterpretations associated with the complexities of the periglacial features themselves, as well as problematic gaps in science and knowledge. The biggest obstacle, however – and this is worth considering in the Canadian context – was that most politicians defined their positions on the basis of ideology, not science. Reconnecting policy with science became a huge problem. Here is how Taillant described it:

> Non-glacier experts formed their own conceptualization of what they understood was a glacier, a rock glacier, or a periglacial environment and forced their personal conceptualization to fit their political objectives. For many, this was more about containing mega-mining or opposing the incumbent administration than it was about protecting glaciers. Many times, the concepts used were incorrect or misinterpreted, but they served the purpose of building a platform upon which to defend a conservationist approach to glacier protection – which is what prevailed in the end, much to the dislike of some members of the scientific community and especially to the displeasure of mining interests.

Taillant goes on to outline the main areas of debate in Argentina's Congress related to the revamping of the vetoed glacier bill. The one perhaps most relevant in the Canadian context centred on whether or not periglacial environments as well as glaciers should receive protection, and if so, which features needed to be inventoried and by whom; what activities should be prohibited in glacier areas or periglacial environments; whether and how activities found currently in such areas should be stopped, redesigned or relocated; and finally, vigorous debate over who should implement the law. It is important to note that in these discussions not even glacier scientists and the scientific community as a whole were able to come to common and conclusive agreement. This became even more complicated when Barrick Gold brought its own scientists into the debate. Their experts wanted the argument to turn on whether or not glacial and periglacial systems were in natural equilibrium, an equilibrium they argued related to larger hydrological processes that were beyond the control of mining and other resource interests, which need to operate in places where, unfortunately, these features are frequently found. On this matter, however, there was scientific consensus. As a consequence of human-caused warming, glaciers and periglacial features are changing and melting, and as a result more water is being released than would have been were it not for anthropogenic influences. It was held that even without adequate research into the precise nature of these dynamics, it was impossible for glaciers not to melt under added warming pressures. In other words, humans, by reason of our needs and numbers, had altered whatever hydrological equilibrium had been thought to exist before, by changing the composition of our planet's atmosphere. The argument that glaciers and periglacial features simply receive and release equal amounts of water was dismissed as baseless. Glaciers and periglacial ice are in essence "water in the bank" for the future, and robbing that bank, it was held, should be outlawed.

That was not the end of the debate, however. Even though

the scientific community agreed on the fundamental concepts related to the role of snow and ice in the perpetual reliability of year-round stream flow, the distortion, cherry-picking and politicization of the science drove many experts to their wits' end, which is probably something we should expect in Canada too if glacier protection becomes a political issue as rapid hydro-climatic change alters notions of water security in the future. At the time of this writing at least, we should likely expect the mining sector in Canada, given their record and reputation abroad, to do what they appear to have done to prevent the passage of glacier protection laws in Argentina and Chile: namely put out enough false or misleading evidence based on distortion of information to confuse the debate for everyone. This is a well-established strategy, as we have already seen repeatedly in the tobacco wars and with the asbestos and climate change debates. The results are almost predictable. In the end in Argentina, lines essentially formed around pro- and anti-mining sentiments. Meanwhile the provinces went ahead with their own glacier legislation. Why? Because they believed that if they acted, the federal legislation wouldn't pass and the provinces would be left in the driver's seat.

Taillant reports that during this period of negotiation, several provincial governors flew to Toronto, ostensibly to meet with Peter Munk and his Barrick executives. Even the president of Argentina apparently took time to meet with Munk when she was in Toronto for the G20 conference in 2010. According to Taillant, model elements of provincial glacier protection bills were drafted during these visits. Laws were then passed in pro-mining provinces, often without legislative discussion. This did nothing to quell the larger debate, however. A very public fight pitted the two sides in televised clashes related to the intention of the federal bill. Then, in the midst of what was becoming a national quarrel, there was yet another unexpected breakthrough. A compromise in the wording of the bill was put before federal lawmakers, and to everyone's surprise, on July 14, 2010, the revised bill passed. After a 14-hour debate during which there were 39 speeches in the lower house in Buenos Aires – in the midst of which the president sent a message that she would not veto the bill again if it were passed – the overall legislation was approved, and in a second vote the specific articles of the bill itself were approved. While mining interests hoped the president would in the end veto the bill, she remained true to her promise. Though it got a rough ride and the vote was far from unanimous, the bill was passed by the Argentine Senate a month later. And so it was that on the morning of September 30, 2010, the world again had its first national glacier protection law.

As already noted, however, it is one thing to pass a law, another to keep it in force, and still another to implement it. Barrick Gold was not going to give up just because a glacier protection law had been passed by both houses of Congress. Three days after the statute was enacted, Barrick obtained exemption from it by way of a Circuit Court ruling in their favour. A year would pass before the Supreme Court of Argentina would overrule the injunction granted to Barrick.

There were obstacles to implementation as well, the most significant being the difficult task of completing a national glacier survey. The legislation called on IANIGLA, the national glacier institute, to undertake a survey within five years. IANIGLA did not think this was possible. Neither could the 180-day timeline for "priority glacier inventories" in mining areas be met. One of the problems here was that mining companies controlled roads into areas of mining activity and were not enthusiastic about opening them to researchers undertaking glacial surveys. Another was that provincial governments which supported mining activity withheld information they had about these areas.

On these problems Taillant appears to have taken matters into his own hands. Working from within an NGO – the Center for Human Rights and Environment – Taillant relentlessly searched Google Earth for relevant satellite images, held information sessions in affected communities and published report after report of his own and others' independent findings. As the saying goes, however, no good deed goes unpunished. Taillant apparently irritated a great many people on all sides of the issue. While becoming an uncompromising one-man revolt outside of formal authority may have been necessary to break the policy logjam in Argentina, Taillant appeared to have missed the point that despite the passage of the legislation, he and others who pushed the glacier protection law were unable to positively stimulate the will to implement it. Some of his activities likely had the opposite effect in deepening resistance to implementation of the law.

That said, it is hard to argue that Taillant didn't believe in the cause, and certainly he had reason to keep on fighting. He himself offers evidence that provincial governments would lie outright about their commitment to implement the law if pressed the wrong way by civil society organizations. Nor did the law appear to change the view of the mining sector, which continued to see ice that was likely to melt anyway as a problem to be resolved rather than a resource to be protected. It also remained difficult to ignore economic arguments raised by the mining companies. Not only that, but 80 per cent of the 180 or more mining projects in San Juan province were at altitudes at or above 3500 metres in areas where conflicts over protection of glaciers and periglacial features would be difficult if not impossible to avoid.

That these matters were not going away but were in fact simmering under the surface became clearly evident by way of the politics associated with the preliminary findings of the glacier survey. In a manner not inconsistent with what a President Trump might have done, politics intervened in the work of the initial survey. It was classic, really. If you don't get the results you want from a glacier survey, then fire the expert in charge and hire someone you trust, not to give you the facts but to tell you what you want to hear. The result, of course, will be a sham survey which would allow mining companies to carry on as they had in the past. But this situation too was gradually overcome.

At the time Taillant published *Glaciers*, the argument over why a glacier law was or wasn't needed was still continuing vigorously in both Chile and Argentina. The real question, however, related to how much environmental damage and hydrological disturbance society is willing to accept from mining. Taillant put into relief two approaches to answering that question. Argentina favoured full environmental and hydrological protection, while Chile appears to have chosen a measure of protection for glaciers but also accommodation of mining interests. In the midst of all this, cryospheric warriors like Taillant continued to question whether we should be trading glaciers for gold. For his part, Peter Munk seemed to be having his own problems over the matter, in the form of a legal siege by shareholders contending that Barrick Gold had misinformed them about the environmental costs of the Pascua Lama mine.

Meanwhile the wheels of government bureaucracy continued to turn. After all, a law is a law. It was accepted that glaciers should be protected as a public good and for their water storage and hydrological regulation value. The law was also tied to the now widely recognized fundamental human right to uncontaminated and reliable sources of water. It was held that mining operations should respect that right. It was also understood that we need to think of glaciers in their larger contexts, not just as isolated features. It was important to understand them not just for their hydrological contributions but also for the way their presence shapes surrounding ecosystems by defining climate, something we clearly need to do in Canada. There is also great value in monitoring glaciers and periglacial features to better understand landform evolution and risks associated with landform stability. The national glacier survey in Argentina undertook all of these tasks. That was in 2015, which brings us back to the INARCH workshop I was attending in Portillo, Chile, and two of the presentations offered there on the current status of formal scientific efforts to monitor the state of glaciers in both Chile and Argentina.

Glaciologist Dr. Gino Casassa began his presentation on glacier monitoring in Chile by explaining that in that country water rights can be bought and sold by the government. (As has happened in parts of western Canada, the state has sold more rights to water than there is water.) Long-term drought and projected dry conditions in the future, Casassa noted, will exacerbate an already tenuous water-security situation. Though he didn't name the companies, Casassa went on to explain that Canadian mining interests had caused a serious crisis. One of the outcomes of this crisis was a formal glacier survey which revealed there were 24,114 glaciers in Chile, covering some 23,641 square kilometres.

Casassa then showed how the melting of ice and consequent darkening of the solar reflectivity of mountain terrain influenced the evolution of landforms. He illustrated how debris-free glaciers become debris-covered glaciers as the mountains around them erode, and how debris-covered glaciers become rock glaciers. Debris-free glaciers, Casassa noted, contribute 60 per cent of the water produced by these combined features, rock glaciers far less. Diminished precipitation and accelerated warming, however, are affecting all glacial and periglacial features. Glacier lake outburst floods have become a serious risk in parts of the country as a result of rapid glacier melt and growing instability of some glacial systems.

Casassa then dropped a bomb. The survey had revealed that, in fact, glaciers in Chile are retreating at the fastest rate in the world, which, in addition to less reliable snowfall, was having a huge impact on Chile's ski areas, including the one where our conference was being held. More urgently, however, prolonged drought and warming impacts on ice and snow were creating a national water-security crisis in Chile. The country was looking everywhere for water, including the water that holds rocks together in features like rock glaciers. Only in very dry places, Casassa concluded, would knowing how much water existed be so critical.

Mariano Masiokas of the Instituto Argentino de Nivología, Glaciología y Ciencias Ambientales – the IANIGLA I had read so much about in *Glaciers* – then presented on the current status of Argentina's first national glacier inventory. He noted first that this was the inventory mandated by National Law 26639, passed in 2010, which provides for minimum standards for the preservation of glaciers and periglacial environments. The survey, Masiokas explained, was organized around three levels of increasing complexity. Level One consisted of the identification, mapping and characterization of all clean-ice glaciers, debris-covered glaciers, snowfields and rock glaciers, using satellite images. The Level One results, published in May 2018, established that there are 16,078 ice bodies in Argentina between the latitudes of 20° and 56° S, covering some 5768 square kilometres. In the inventory's next iteration, levels of legislative protection would be matched with the character and location of each ice mass as determined in that stage of the survey.

Masiokas's presentation put our understanding of our Canadian glaciers and periglacial features into interesting relief. We know comparatively little about our own ice. There are ten staff involved with Argentina's National Glacier Survey. In Canada we have only two people working on glaciers for the Geological Survey, neither of whom is involved in any form of national surveying. If a national glacier survey is held to be in the national interest of Argentina, wouldn't it be in the national interest of Canada to inventory, monitor and protect our glaciers and related periglacial features, especially given the alarming speed at which hydro-climatic change is advancing both in the Arctic and in Canada's western mountains? Come to that, why wouldn't humanity protect glaciers in this way globally? It was worth coming all the way to Chile just for these two presentations.

So, can a case be made for Canada to enact a glacier protection law? Let's compare some numbers. The Canada Water Act was passed in 1985 and has not been modified in any significant way since, which at the time of this writing was 34 years ago. The Bible makes 41 references to ice. The Canada Water Act makes no mention whatsoever of ice or glaciers or protection thereof.

According to the World Glacier Monitoring Service there are roughly 160,000 glaciers in the world. Some 55,000 of these are found in the Canadian Cordillera. About 9,600 more can be found in the Canadian Arctic, for a total of about 64,600. This suggests that Canada possesses some 40 per cent of the world's glaciers.

As mentioned earlier, there are an estimated 16,078 glaciers in Argentina. Thus Canada's 64,600 would amount to about four times as many. Yet even though Argentina possesses only a fraction of the number of glaciers that exist in Canada, they now have a glacier protection law. Shouldn't we?

The answer to that question is: probably not. Constitutionally, most water quality and resource matters are under the jurisdiction of individual provinces and territories. The only Canadian province that presently has glaciers and mining on or near them is British Columbia. If such legislation should come into existence, it should be there. But the matter of a glacier protection law invites a larger question. Should we have better integration of watershed protection legislation tied to resource development in each province and territory as well as federally? Do we need a national water commission? Do we need a national water-security research centre? Yes, absolutely.

NINE

AS THE WORLD BURNS

Whenever the conversation turns to national strategies for managing water-related climate impacts – while at the same time protecting and bolstering regional natural capital, redesigning cities and preparing citizens for the very different world we are bringing into existence – many are compelled to look again and again to China as a model for imagining and then creating a sustainable future for all of humanity. Why? Because as China goes, so goes the world.

China is the world's biggest emitter of greenhouse gases. As I reported in my 2018 book, *Quenching the Dragon,* the burning of coal is the main reason why China is home to 16 of the world's 20 most polluted cities. Also, according to 2010 projections, in 25 years China will emit two times more CO_2 than all of the OECD countries combined. The country will also inherit serious climate change impacts, partly as a result of its own emissions.

In that book I also reported that the total annual cost of environmental decline in China had already been estimated at $111.6-billion, a staggering 8 per cent of China's GDP even then. Gross domestic product, however, can hardly be considered a useful measure in determining the state of the environment anywhere. The question now is not whether pollution generated by China is costing the rest of the world, but how much.

At the moment we cannot answer that question. All we know is that we are failing to meet emissions reductions targets globally, at huge potential cost to the future, and that there are no penalties for such failure.

It should surprise no one that the combined heat generated by big emitters like China and the U.S. is now seen to be having deadly direct and indirect impacts on neighbouring countries whose total emissions they hold to be relatively inconsequential. As I will report in the next chapter, many people in the northern hemisphere woke up to this realization in the summer of 2018, when flooding was followed by heat waves, drought and wildfire for the fifth year in a row in Canada and across much of the hemisphere. The total cost to the rest of the world of China's pollution? Inestimable. Or is it?

Why does that matter? The dangerous economic, social and environmental risks our civilization takes in not putting natural system restoration on par with economic growth as a political priority are put into relief by the terms and conditions of the Paris Agreement on Climate Change, to which most nations of the world, including China, are signatories. The Paris Agreement was constructed in such a way that its central goal was to both limit warming to "well below" 2°C and "pursue efforts" to limit warming to 1.5°C. It is a given, however, that since the amount of carbon emitted cannot be known exactly, the response of the Earth System to human-generated greenhouse emissions cannot be known precisely either. Consequently, temperature change projections are based on probabilities.

A fair reading of the Paris Agreement would be that the targets agreed upon would have a 66 per cent probability of limiting warming to 1.5°C and a 90 per cent probability of not exceeding 2°C. To meet the 2°C limit even within the 66 per cent probability range would require rapid cuts in greenhouse emissions to near zero by 2050. How close are we to meeting that target?

While the comprehensive and global Paris Agreement is truly an historic milestone, there are no clear penalties for the failure of any country to actually undertake its pledged actions. While the agreement allows for examination of the risks of damage and displacement, it does not allow attribution of blame, suggestions of liability or recommendations for compensation. It is important to note also that the negotiations required to arrive even at this level of agreement did not go smoothly. It took from 1992 to 2015 to get to where we are now.

The country pledges, called Nationally Determined Contributions, go nowhere near far enough to achieve the "well below" 2°C goal. The gap between promise and practice continues to grow: the discrepancy between the pledges and a 2°C outcome are roughly equivalent to the combined emissions of China and the United States in 2017. With President Trump reversing America's commitment to reducing greenhouse gas emissions, this gap is set to become even wider.

Because global greenhouse emissions are increasing rather than being reduced, the expectation that warming will be limited to 1.5°C or "well below" 2°C is already held to be unreasonable. According to many climate scientists we may be looking now at temperature increases of 3°C or perhaps as high as 4°C within the next 75 years or, as suggested by what we saw in the northern hemisphere in summer 2018, even sooner.

While it is technologically possible to reduce emissions to near zero, the only way this target could be achieved would be to put the eradication of greenhouse gas emissions at the same level of importance as economic growth. Understood more generally this would suggest that for China to succeed – for any country to succeed, including Canada – complete restoration of Earth System stability, not just elimination of emissions, would have to rank equally with economic growth. No one who understands the problem needs to be told that this is not going to be easy anywhere, even here in Canada.

In assessing China's progress in the direction of sustainability, it is useful to consult sources such as the International Institute for Sustainable Development. In a July 2018 article called "China's War on Pollution – And What Comes Next," for example, the IISD correctly acknowledges that what China does is key to the advancement of sustainable development globally. The article further observes that China has set ambitious targets aimed at meeting the UN's 2030 global sustainable development agenda, including a strong commitment to the Paris Climate Accord. These actions, the IISD notes, fall under an evolving concept which China is sharing with the world with the aim of creating an "ecological civilization." Such policies had already been somewhat overshadowed in October 2017, however, when President Xi Jinping declared at the 19th Chinese National Congress that his country was about to abandon self-isolation to instead pursue the goal of replacing western liberal democracy with China's unique brand of market socialism globally. Still, despite receiving less attention abroad, the ecological civilization initiative is of critical significance not just to China but to the rest of the world. Its policies include a comprehensive package of national environmental protection measures and climate change mitigation actions, including enhanced forest conservation; an increase in marine protected areas; the acceleration of industrial planning to reduce air and water pollution; and a commitment to strengthening national environmental monitoring and regulatory enforcement protocols.

In May of 2018 President Xi further confirmed that the concept of ecological civilization would be at the centre of China's contribution to the UN's 2030 *Transforming Our World* sustainable development agenda. It was widely reported at the time that Xi went so far as to note that "lucid water and lush mountains are invaluable assets," a comment that could be interpreted as a nod in the direction of valuing and restoring the country's natural capital. A clear-sighted analysis of the pillars of the ecological civilization promise reveals, however, that the order of China's national priorities has not changed. The three pillars of sustainable development remain economic growth first, social development second, and only then improvements in the environment.

The cynic might ask how it would be possible to address the problems growth has created by simply stimulating more growth, but everything in China is more complicated than it might appear from the outside. Any analysis of the ecological civilization idea must take into account China's long and remarkable history and the depth of its culture in the context of its current governance model. So here we have it: China will lead the world toward a sustainable future through the example of its own economic, social and environmental development. There is little doubt in senior Chinese political circles that this goal is achievable, and quickly. Outside observers are not so sure. In its strategic efforts to forcefully assert its presence on the world stage, China is already actively seeking to recast international norms and institutions in its own image. Rhetoric notwithstanding, the fact that China is an illiberal state seeking leadership in what at least at the moment is a liberal world order cannot be ignored.

According to the IISD article noted earlier, the most important forum for promoting the ecological civilization concept is the annual Eco Forum Global conference, held in Guiyang, in southern China. If one attended that conference, however, it might be easy to conclude that its proceedings focused more on economic growth and corporate opportunity than on protecting, restoring and sustaining natural Earth System function. The Chinese invariably exhibit a great deal of confidence at these forums. So much so, that visitors often feel they have come up against another of China's Great Walls: its legendary and rapidly growing arrogance concerning its place on the world stage, an arrogance that has grown since the election of Donald Trump as U.S. president. America's clumsy and embarrassing decision to exit from the centre of that stage of its own accord has enabled China to rush in and play its part without yet learning the lines of the major actor who must play that role. When I myself witnessed this famous arrogance of China's in Guiyang and later in another global forum in Beijing, I couldn't understand it. Now I think that, to some extent at least, I do.

In *The Man Who Loved China*, his biography of British scientist and scholar Joseph Needham, Simon Winchester put this arrogance into relief, noting that while the Chinese are not as racially homogeneous as the Japanese or the Koreans, they nevertheless remain steadfastly disinclined to dilution or change. Instead they favour periods of long-term ethnic stability. This does not mean, however – as the Tibetans have discovered – that the Chinese are above a cautious but relentless expansion of imposed ethnic homogeneity within their own country and where possible in surrounding regions and elsewhere. President Xi made this clear in a 2014 speech in which he observed that China is now capable of "constructing international playgrounds" and "creating the rules" of the games played on them, something China's tourism and trading partners need to take into full account in the context of the quiet but rapid expansion of Chinese ownership of property and businesses in countries like Canada.

But within the desire of the Chinese to remain distinctly Chinese there is something else that resists change, or at least change that does not originate within the culture itself. That something, which resides just below the surface of national sensibilities, can only be described as an attitude, a Chinese state of mind that outsiders – which as Winchester points out includes everyone who is not Chinese – may sometimes find "infuriating and insufferable." It is an attitude born of the long-term achievement of Chinese culture. The legacy associated with the country's centuries-long record of innovation and technological advancement has, in Winchester's estimation, led to an ingrained attitude of "ineluctable and self-knowing superiority," a superiority founded on the longevity and persistence of Chinese endeavour throughout history. This sense of superiority is buoyed by a phenomenal list of inventions and cultural improvements which, since antiquity, have been making life better, easier and more civilized, at least for an elite, for most of the history of China in comparison with the rest of the world. Whether by inventing the compass, printing paper, the wheelbarrow or the suspension bridge, the Chinese have always been ahead. With some 15 major inventions per century throughout its long history, China feels justified in its sense of self-satisfaction and superiority.

Winchester notes, however, that China's natural smugness

may have at one point led to self-congratulatory complacency – even hubris – which in part led to the backwardness and poverty from which the nation is only now emerging. But in its re-emergence, that sense of superiority is still there: "that peculiarly and infuriatingly *Chinese* sense of self-certainty, of an unshakable confidence" as Winchester calls it, a trust in its capacity for innovation and invention that will allow China to take its rightful place at the centre of the world. The question then becomes – and this is critical to the concept of an ecological civilization – whether, how quickly and how competently China can capitalize on its earlier historical legacy. In answer to this question, Winchester draws his readers' attention to a simple sign outside Jiuquan, a modern and gleaming new town in the midst of the Gobi Desert that is now one of China's most important satellite launch sites. The sign, in both Chinese and English, simply says: "Without Haste. Without Fear. We Conquer the World."

After 5,000 years of patient waiting, Winchester observes, this may at last be China's appointed time. The Chinese certainly seem to think so, and with the u.s. inexplicably surrendering its place on the world stage as noted earlier, China's timing may well be excellent. But there is one big fat fly in the shark-fin soup: the long-term deterioration of air and water quality and other environmental conditions throughout most of China. Hundreds of thousands of premature deaths and incidents of serious respiratory illness in China are attributed to exposure to industrial air pollution. The country's rivers are in much the same condition as its air. The Yellow River, which provides water to millions of people in northern China, is now so badly polluted that 85 per cent of it has been declared unsafe for drinking.

The IISD article mentioned earlier acknowledges that China has declared "war" on air, water and land pollution and is developing more stringent regulations, deploying expanded air and freshwater monitoring systems with the goal of toughening pollution enforcement measures. The article notes also that China is planning the largest expansion of urban planning in history. The object of this, we are told, is to embed clean, low-carbon pathways into urban planning procedures. The IISD goes on to explain that the dividends of these processes will be quantified by way of measurable improvements in air quality, particularly in cities.

The progression of this command-and-control process is going to be interesting to follow. The notion that making where you live habitable should be characterized as a "war" notwithstanding, there are fears that China's economic ambition and arrogance with respect to its own organizational and inventive capacities could be its downfall. In this, observers worry that the country has placed far too much emphasis on technological innovation and invention in this so-called war, and not enough emphasis or trust in the need for restoration of lost or damaged natural systems. This perspective puts into relief what I consider a consistent flaw in many geopolitical projections and many foreign policy analyses. Few, if any, appear to take planetary health into account, and even when they do, it almost always takes a back seat to economic growth, which – whether stated or not – is the first and highest priority.

Authoritarian governments have historically tended to place little emphasis on protecting natural system function until it is too late. Parts of China are already degraded, their natural capital nearly exhausted. Given that some of these regions are barely habitable now, it is not inconceivable that some will become almost uninhabitable for at least parts of the year, if not indefinitely, in a warmer, hotter and more variable climate. This will undoubtedly mean further migration within the country, voluntary or otherwise, into areas of remaining ecological resilience. The fear is that in a changing climate, the failure to sustain natural system function will cost countries and regions – and the rest of the world – more than we can afford. At the conclusion of "China's War on Pollution – And What Comes Next," IISD former president and CEO Scott Vaughan underscored the seriousness of this threat. The findings of IISD's work in Canada, he noted, affirm disturbing findings that the stock of natural capital globally is declining at an alarming rate: by as much as 25 to 30 per cent in the past decade alone. At this rate of loss of natural resilience and absorptive capacity, it will not be long before most of the world is exposed to the full force of steplike changes in Earth System function that include higher temperatures, changes in precipitation, and more life-threatening and infrastructure-damaging extreme weather events. This has to be seen as a huge threat to China and its aspirations for renewal economically and socially as well as for the appointment with destiny it hopes to keep with its spotlight reappearance on the world stage.

If that appointment is to be kept, and for it to have any lasting meaning, China must do two things while boosting its economy: it must help the world stabilize the composition of the Earth's atmosphere, and at the same time aggressively restore its own lost and damaged natural system function – starting with water – without exporting that damage elsewhere. In this, China will need all the help the rest of the world can provide. At the same time, the rest of the world should do all it can to encourage China and appeal to its proud history of invention and brilliant innovation. China has just had its industrial revolution. The country is now viewed by many as the world's largest contamination site. We – that is to say China and the rest of humanity – have to help clean up the mess before restoration or rehabilitation of lost or damaged natural function in that critical part of the world can even begin. The world needs to give China the technology available today, and then see what they do with it and what they do to advance it. Why? Because in terms of human health and well-being, the stability of the Earth System and fundamental planetary health, as China goes, so goes the world.

We are not without solutions. We know what we have to do. In order to deal with the damage we have done to the world, we need to remake our society, just as China is trying to do. We can do that under the aegis of a concept that has emerged from the UN's 2030 *Transforming Our World* global sustainable development agenda: the idea of a "restoration imperative." Such an imperative would demand that we restore a global commitment to human dignity and fundamental rights, and commit to a vision for the future of humanity and the planet. But most urgently such an imperative must become an immediately effective vehicle for the protection and rapid restoration of critical natural system function so that we can return to balance in the world and step back from the climate brink.

Let restoration be our imperative. We can create an ecological civilization. We have the InterAction Council's Dublin Charter for One Health to guide us. We also have, at last, a comprehensive definition of what sustainable development really means and embraces, and a firm timetable for achieving sustainability globally.

Sustainable development as defined by the 2030 *Transforming*

Our World agenda makes it clear that unless we all take the same common goals seriously and implement meaningful and measurable actions at the national level in every country in the world, now, we will not achieve sustainability globally. This means there can be no laggards, particularly in the developed world. It also means the world cannot afford to leave anyone behind.

We also can't afford the world we have created and live in now. We have to stabilize and restore that world and then start down a new path toward enduring sustainability, toward an ecological civilization. To do that – whether we like it or not – the rest of the world needs China.

TEN

LEARNING FROM THE BURNING: THE SUMMER OF 2018

In my work, the goal – always – is to build bridges between scientific research network outcomes and public policy on water matters and water-related climate change impacts. I work to find ways to identify and then reduce the effects the growing global water crisis and water-related climate disasters are having on the stability of particularly fragile member states of the UN. I also work to identify and act on threats that climate change is already posing to peace and security in many regions and on involuntary human migration that may lead to conflict within and among UN member nations. But it is no longer just fragile states that are vulnerable to these impacts.

The foundation of my work is science. It is my belief that because of the rigour of its practice and the way its high ideals bring the world together in the common interest of improving our collective knowledge, the scientific method is arguably the most powerful tool humanity has in our endless pursuit of the meaning of our existence. I am of the view that the best of science understanding goes far beyond the influence of intellect on rational being; it sets fire to the human imagination in ways that give greater understanding and deeper significance to where and how we live our lives. Good science is not just the sharing of knowledge about the world; it is a candle we light when we want to see and be warmed by the truth.

There has probably never been a time in history when making science understandable to a vastly diverse public has been more important. We live in an era when we are so awash in data and information that we often don't always know how to give it the fullest meaning and value until the opportunity to do so is long past. We are also at a bottleneck in the evolutionary history of our species where failing to understand and act appropriately on what we know could have devastating impacts on future generations, and potentially catastrophic effects on Earth System function for the rest of time.

It is my hope that appealing to leadership at all levels of society will ultimately contribute to making sustainability possible, not just in Canada but around the world, by putting reliable evidence before leaders and those who would become leaders. I further hope to demonstrate the crucial importance of convincing others of the urgency of immediate action on the global water crisis and accelerating hydro-climatic change. It is a hateful job. I am the deliverer of bad news that no one wants to hear.

On the matter of urgency I continue to be drawn back to a comment made by a famous Canadian fisheries biologist and world expert on salmon, Kim Hyatt, at a conference on protection of environmental stream flow in precious, productive aquatic ecosystems in British Columbia's Okanagan basin. The premise that keeps Hyatt going and now keeps me going in this work is this: I would rather be afraid now so that our grandchildren don't have to be afraid in the future. Indigenous Elders and keepers of ancient wisdom around the world repeatedly remind us that our work in our time is all about future generations. The Indigenous commitment to seven generations forward is not a political stump speech. It is an ethic we must all share if we want a worthwhile future, here or anywhere.

In Canada, but particularly in provinces like British Columbia, huge efforts are being made at local, regional and provincial levels to make true sustainability, founded on historical reconciliation with one another as peoples, into a basis for restoring and rehabilitating a deteriorating world. Wisely, many of these efforts revolve around issues related to water. I am often invited to give evidence relating to such matters that aims not only to underscore the importance and urgency of such work but also to inspire the continued efforts of the many, many dedicated people in Canada and around the world who are wholly committed to a positive vision of the future.

By the autumn of 2018 it was no longer necessary to tell anyone involved in this work why what they were doing was more important than ever, and why communicating the value of their work to decision-makers had become all the more urgent. The Intergovernmental Panel on Climate Change had just put that urgency into staggering relief. The IPCC report released in October of that year stated in very plain language that averting a climate crisis will require an immediate and wholesale reinvention of the global economy within 12 years. It was too much for many, who simply dismissed it. For many climate scientists, however, it wasn't enough.

By its nature, the IPCC is a very conservative organization. The last thing they want to do is overstate the issue and have critics condemn them for doing so without adequate justification. If there is a weakness in their 2018 report, it resides in the scientific method itself. Science is a slow and grinding process.

Peer review takes so long and is so rigorous that the data on which the IPCC made its judgments was already ten years old.

There was nothing inaccurate in the IPCC projections. What they reported was as far as they could agree to go in terms of alarming the world while still being accepted as credible. Some very high-profile researchers, however, including Michael Mann in the U.S. and John Pomeroy in Canada, believe that in its efforts to avoid being too alarmist, the report downplays the economic costs of severe storms and of the involuntary displacement of people due to drought and deadly heat waves.

A recent Parliamentary Budget Office report on the federal Disaster Financial Assistance Arrangements program, which reimburses Canadian provinces and territories for damages resulting from natural disasters, suggests that such an underestimate in the IPCC report may well be the case. The average total annual payments in indexed dollars rose from $54-million for each year from 1970 to 1994, to $291-million from 1995 to 2004, to $410-million annually from 2005 to 2014, an increase of 7.5 times. It is estimated that the total damages to the Canadian economy caused by extreme weather events from 2000 to 2017 exceeded $28-billion.

If you think the damages and costs of extreme weather events have cost Canada dearly, look at the numbers from the more heavily populated U.S. states. The National Centers for Environmental Information reports that their country has sustained 238 weather and climate disasters since 1980 where overall damages/costs reached or exceeded US$1-billion (Consumer Price Index adjustment to 2018). The total cost of these events exceeds $1.5-trillion, yet Mr. Trump continues to deny the climate threat.

It is for reasons like this that others argue it doesn't really matter what the Parliamentary Budget Office, the National Centers for Environmental Information, the IPCC – or anyone else, for that matter – report, because few political leaders take the climate threat seriously anyway. What is new and really dangerous about this, however, is that we now understand that runaway climate change is possible even at 2°C of warming, never mind 3° or 4° as we had previously thought. Moreover, after that 2018 summer many prominent scientists were expressing the belief that, in the northern hemisphere at least, we may already have crossed over an invisible threshold into a new, more volatile climate regime.

So, why the difference in these views? Certainly, part of it resides in the fact that global consensus on climate risks is always difficult, and therefore the IPCC most often errs on the conservative side. There is also the problem with global averages. All you have to do to understand that problem is think about how representative an evaluation of the water security of Canada would be if it were based only on averaging the water availability in the Great Lakes basin with that of the prairies and of the Okanagan Valley in British Columbia. In a continental context that average would be almost meaningless.

In the same way that national averages obscure what are often huge regional differences, the IPCC projections conceal large differences in the variability of effects at the sub-planetary level. The global average does not accurately represent hemispheric and continental climate change effects. In Canada, for example, the Arctic is already far above the 2°C limit to which we need to hold temperature increases to avoid catastrophic warming.

Climate researcher Dr. Mel Reasoner explains that in the Okanagan, for example, you can expect the historical climate to move toward more hot weather and more record hot weather. Reasoner notes that with our current business-as-usual scenario we should expect the Okanagan city of Penticton to have the same summer climate as Karachi, Pakistan, by the end of the century. It should be pointed out that Karachi currently has a population of 14 million. It is highly unlikely that Penticton would ever get that big, but the population growth rate in the Okanagan is the fastest in British Columbia, and warming there is advancing at one and a half times the global rate.

Recent estimates are that even at 2°C of warming globally, we should expect at least 5°C in Canada's western mountains, as much as 9°C on the interior plains, 10°C farther north in our boreal forests, and even greater warming in the Arctic. We know also that we are not going to limit mean warming to 2°C.

So what are we talking about here? First of all it should be noted that all of these effects are related in some way to water. While I have written about this in other books, it is important here to remind ourselves of the climate change effects that are most pronounced in the northern hemisphere at present. I have updated my reporting of these effects from earlier writings where possible.

Land use and land cover changes are only the beginning of the effects human activities are having on the global hydrologic cycle. Our entire Earthly reality is defined by all the ways in which water reacts with nearly every element in the physical world. Change a few parameters that pertain to water, and the world you see out your window becomes different. Some parameters, however, have more influence than others over the nature and function of any given hydro-climatic circumstance. The changing of a single defining parameter – temperature, for example – changes all the other biogeochemical parameters. If our global temperature changes, an entirely new geometry is created around that change. What this causes is the loss of what we call hydrologic stationarity. The most frightening discovery of this still-young century is that this is exactly what is happening. The rate and manner in which water moves through the global hydrological cycle is accelerating. It is happening right before our very eyes – here and everywhere.

The most profound changes to the hydrological cycle relate to how much more water a warmer atmosphere can hold. If you warm the air by 2°C, it can carry as much as 14 per cent more water vapour. Raise the temperature by 4°C and the atmosphere will carry 28 per cent more water vapour, which changes everything. More water vapour in the atmosphere makes storms more powerful, heat waves more intense and drought deeper and more persistent. And that is why recently identified phenomena such as atmospheric rivers demand our fullest attention.

Atmospheric rivers are now seen as the workhorses of water vapour transport. They are responsible for fully 22 per cent of global surface runoff. Because of warmer air, these huge rivers of water vapour aloft are carrying more water and causing flooding of magnitudes we have not witnessed before. In a warmer climate we can expect 40 per cent more activity in atmospheric rivers and a 25 per cent increase in their width and duration.

The power of these extraordinary water extremes, we have found, is tied to the temperature gradient between the mid-latitudes and the poles, a gradient that is being diminished by warming of the global atmosphere. This also affects the jet stream.

Put simply, the northern hemisphere jet stream – often

called the polar jet stream – is made up of the westerly winds created as the world spins inside the loose clothing of its own atmosphere. The behaviour of the jet stream is also tied to the temperature gradient between the mid-latitudes and the poles, a gradient that is now breaking down.

Anyone who has watched a weather forecast on the evening news will have noticed, I am sure, that the jet stream curves around in great meanders. The first sign that something important was changing was that the increase we have been witnessing in recent years in the waviness of the jet stream began bringing cold Arctic air to ever lower latitudes than in the past. People say to me all the time: How can there be climate change when our winters are still so cold? Here is what is happening. In addition to being larger in amplitude, the meanders in the jet stream move eastward more slowly, causing weather patterns also to move more slowly. It is these more persistent weather patterns that result in intensification of the events we are witnessing in which duration is an important factor, such as droughts, floods, heat waves and cold spells.

We see from the altered behaviour of the jet stream that warmer atmospheric temperatures do not automatically translate into warmer weather. In a uniformly warmer and therefore more turbulent atmosphere, both warm and cold fronts end up and persist in places in the mid-latitudes where they were not common in the past. And they oscillate there, often with disastrous effects.

The changing behaviour of the jet stream appears also to be directly linked to the extent of sea ice in the Arctic, where warming is already 3.5°C above the global average. Continuously warming temperatures are leading to more and more sea ice loss, which in turn makes the jet stream even wavier, which brings even more warm air into the Arctic, which leads to more sea ice loss, creating a self-amplifying feedback loop that goes on and on. The Arctic is now warming five to eight times faster than the equatorial region, further altering the temperature gradient between the poles and the tropics.

As a consequence there is also the related threat posed by natural methane and perhaps mercury releases from the warming land into the air and into inland and coastal waters. But thawing permafrost is only one concern. As the Arctic Ocean warms, frozen marine sediments containing methane hydrates and clathrates are also thawing, producing methane gas that has begun to rise to the surface in great bubble plumes which may have a greenhouse effect that can be as much 100 to 200 times that of carbon dioxide.

Alarmingly, the overall combination of snow and ice loss in the northern hemisphere may contribute an additional 50 per cent to the direct global heating effect caused by the addition of CO_2 to the atmosphere. It is also important, especially in the context of recent intensification of wildfires, to understand that there is a direct climatic connection between water and its diametric and symbolic opposite, fire. If you don't have the one, then you get the other. Higher mean annual temperatures, especially in northern Canada, combined with the mismanagement of wildfire over the past century, have resulted in a greatly increased wildfire risk. A slower and wavier jet stream will cause the conditions that exacerbate the wildfire threat to intensify and persist longer in vulnerable places. Based on carbon dioxide increases alone, scientists predict a whopping 75 to 120 per cent increase by the end of the century in the extent of area burned each year.

The combined effect of all of these impacts is the gradual loss of distinct seasonality as we have known it. Nowhere is this more evident than here in the Canadian West. So why aren't we doing anything about all this?

Scientists do not make rash pronouncements. They are especially loath to fall into the trap of pareidolia, the human tendency to look at a random image and see a pattern where none exists. The rapid retreat of northern hemisphere sea ice and the amplification of Arctic warming, for example, have all occurred in only the last ten years, but ten years is not enough time to distinguish the effect of changes in the Arctic system from other, random events in the global climate generally. What I am really saying here is that at the moment, in the Arctic at least, climate disruption is moving faster than science can keep up. There is strong evidence that total Arctic sea ice loss could have an irreversible effect on the stability of the climate of the entire northern hemisphere.

We have known for years that we were approaching this tipping point, but it was held that we likely wouldn't know where it actually was until long after we had crossed it. Many prominent climate scientists fear, in part because of what is happening in the Arctic, that we may have crossed that tipping point during the northern hemisphere summer of 2018.

I worked throughout that summer to massage the tightening knot in my chest created by the accelerating climate disruption we were seeing throughout much of the northern hemisphere. The object of that stress was to find words to describe my growing anxiety about what we were observing, and to turn those words in the direction of positive action.

After examining the evidence I arrived at the view that if it hasn't already happened, the heightened anxiety that I and many others are already experiencing personally about climate disruption is about to become a permanent part of the human condition. We will, all of us, share the anxiety many throughout the northern hemisphere were feeling during that summer as week after week we were unable to leave the smoky room of a rapidly changing climate. Rather than despairing about it, we decided we needed to put this anxiety to work for us and for the future of humanity. We can do that by urgently focusing on protecting, restoring and rehabilitating natural system function and resilience where we have the most power to effect change: in the world immediately around us.

The first question the reader might ask is: How can the claim be made that climate disruption is an immediate enough threat to warrant extraordinary efforts – efforts far beyond those we are presently committed to – to restore natural system function? The first exhibits I would like to put before the court of the reader's opinion are advances in the attribution of human influence on the intensity, frequency and duration of extreme weather events globally.

While it had been conventional wisdom among scientists for decades that no single event could be attributed specifically to climate change, the latest research indicates this may no longer be the case. A major report published by the prestigious U.S. National Academies of Science in March 2016 clearly states that science has advanced to the point where, in certain instances, anthropogenic attribution may be possible.

The principal reason why we are currently incapable of accurately attributing any extreme weather events to anthropogenic causes is not because such events are not real. It is because our models are still in development or because the long-term data we have is limited. It is also true that in many cases we don't even know enough about some forms of extreme weather to make reliable judgments on the degree to which humans could influence their intensity and duration.

That said, the science behind the attribution of extreme weather events to anthropogenic influences has advanced a great deal in recent years and is evolving. Depending on the nature of the event, the framing of questions about attribution and the availability of relevant data, tentative assessments can now be cautiously put forward.

Generally we know a fair amount about extreme heat and cold events, largely because of the quality and length of the observational record. We also know a good deal about droughts and extreme rainfall. We know less about extreme snow and ice storms, however, and even less about tropical and extra-tropical cycles. Nor are we very good at predicting or assessing the behaviour of extreme convection events like tornadoes.

Attribution of extreme weather events, moreover, is a rapidly developing science, an exploration of causality. From the outset it is recognized that extreme weather has multiple causes. Always. Large-scale meteorological patterns are complex. They involve big-system dynamics that include the Atlantic Multi-decadal Oscillation, the El Niño Southern Oscillation and the related Pacific Decadal Oscillation. This may not be as deep as it sounds, though. With reliable long-term data, attribution is, under certain circumstances, becoming possible.

All attribution statements remain highly conditional, and attribution is harder to estimate in hydrological events, but with the right data it can be done. It has been demonstrated, for example, that human impacts on the climate system made the flooding in Colorado in 2013 – the same year as the flood disaster in southern Alberta – twice as likely and made rainfall 30 per cent more violent. Studies conducted under the World

Climate Research Programme on 35 of the 63 extreme seasonal events between 2011 and 2013 found, after 400 model simulations, that human influence was clear on all of the heat waves during that period. Evidence suggests that human impacts made India's 2015 heat wave 30 times more likely. In neighbouring Pakistan, recent heat waves were up to 1,000 times more probable because of lost or damaged natural system capacity to absorb the effects of extreme conditions. It is estimated that human impacts on the climate system made the heavy rainfall associated with Hurricane Harvey 19 per cent to 38 per cent more likely. It will be interesting to see the extent to which the ferocity of future hurricanes will be attributed to anthropogenic causes.

It is now held that the sweltering northern hemisphere summer of 2018 would not have been half as likely were it not for human-induced global warming. Recent research done in part by Canada's own Pacific Climate Impacts Consortium demonstrates that during the current decade the combined effect of anthropogenic and natural climate forcing is estimated to have made extreme fire risk events in western Canada 1.5 to 6 times more likely compared to a climate unaffected by human influence.

What will the growing capacity to attribute the effects of human impacts on extreme weather events mean in terms of financial liability? We don't know yet. But what we do know is that here's no better time than after the northern hemisphere summer of 2018 to talk about liability with respect to climate disruption, for during that summer we began to see exactly what inaction is going to continue to cost us, not just economically but in terms of human duress and suffering, if the climate threat is not addressed. As a consequence of that summer, municipalities in British Columbia contemplated class action lawsuits against the world's largest fossil fuel producers to recover damages from yet another devastating province-wide fire season. It will not be easy. While the Paris Agreement allows for examination of the risks of damage and displacement due to climate disruption, it does not permit attribution of blame, allegations of liability or recommendations for compensation. It is a big question, and it is becoming a global issue. Who is going to pay and how much?

While the scientifically comprehensive and worldwide Paris Agreement is truly an historic milestone, there are no clear penalties for the failure of any country to undertake its pledged actions. Because global greenhouse emissions are increasing rather than diminishing, limiting warming to 1.5°C or "well below" 2°C is already held to be an unreasonable expectation. According to many climate scientists we may be looking now at temperature increases of 3°C or perhaps as high as 4°C within the next 75 years or, as suggested by what we saw in the northern hemisphere in 2018, even sooner.

The next exhibits I would like to put before the court of the reader's opinion relate to whether or not there is evidence enough to suggest, based on the summer of 2018, that we have crossed an invisible threshold into a new climate regime.

The turning point in our observations was the sunrise on the morning of Friday, August 17, 2018. Over much of western Canada it was surprisingly smoky in an apocalyptic way that was eerily reminiscent of some of the sci-fi doomsday thrillers of the 1980s and '90s. It had already been a hot and deadly summer across the entire northern hemisphere.

For us in the mountain West, that summer had begun with the rapid melt of near-record snowpacks. This unusually fast and early snowmelt caused record flooding in parts of British Columbia and exceptionally high flows in rivers draining the Canadian Rockies.

Then came the real heat waves, and not just here. Japan experienced its highest temperatures ever: 41°C in some cities, with humidex readings as high as 48°C. Record temperatures killed at least 29 in South Korea. The June heat wave in eastern Canada resulted in some 93 deaths in Quebec. Heat was matched by lack of rain in the Canadian West, with the worst drought in 50 years in parts of the southern Prairies.

Next came the wildfires and the endless smoke that spread across five provinces and two territories for weeks on end. But this too was not just in Canada; it happened right across Europe and Asia, with floods, droughts and fires everywhere. The impossible seemed to be happening all at once and all around the world. Rapidly moving wildfires in Greece killed 91 people in July. All of Europe was fighting wildfires, from Portugal to the Swedish tundra north of the Arctic Circle and across the normally soggy Welsh mountains. Down under, Australia was fighting unprecedented winter bushfires. California had its worst fire season ever, even including a "fire tornado." At the same time, France and Italy were fighting deadly floods.

Meanwhile, the Americans in the eastern and upper midwestern states were also battling floods, as were West Africans, Colombians and Venezuelans. By August the most severe monsoon rains in over a century had killed over 400 in India and displaced over 800,000 people from their homes.

At the same time, the thickest depth of multi-year sea ice north of Greenland was melting and breaking up. It was thought that this old, thick ice would last the longest in the Arctic Ocean, but that turned out not to be the case.

Also in 2018 we once again saw a slower and wavier jet stream cause the wildfire threat to intensify and persist longer in vulnerable places. As a result, fires became bigger, hotter and faster. And just as scientists had predicted, methane releases accelerated due to thawing permafrost in the Arctic. We now have forest fires fuelled by the very methane they release.

We also observed, as noted in chapter 3, that megafires can generate enough energy to enable what are called pyrocumulonimbus clouds to pierce the stratosphere. Ash and debris carried that far aloft circulate in the upper atmosphere in the same way as from big volcano eruptions, and with the same result, albeit with lesser potential short-term influence on the global climate.

We know the effects on us physically and what that will cost in terms of healthcare. But what do we know of the mental-health impacts of living under a smoky sky for weeks on end, summer after summer? How will people deal with prolonged periods when they can barely see across the street, their eyes and throats burn, and the air is hard to breathe? We saw in summer 2018 what it is like to live in many parts of China today, where people don't want to go outside; it is unhealthy to ride your bike or work in your garden; and outdoor public events get cancelled because of poor air quality. Wildfires are happening more often around the world and have come to be as costly to our society as hurricanes and tornadoes. Air quality from fires is a worldwide issue. It is estimated that 330,000 people globally will lose their lives every year from the aftereffects of wildfire.

During the summer that wasn't, with reports of over 560 forest fires burning in B.C., thousands of people under evacuation notices and smoke affecting most of western Canada and the northern U.S., many were thinking about where we are headed and just how fast we might get there.

Prominent climate scientists are now saying their worst fears have been realized. The feedbacks they cautiously projected might further accelerate climate impacts over a longer time span now appear to be kicking in far earlier than expected. The impossible seemed to be happening all at once around the world. And it kept on coming. As mentioned, California had its most severe fire season ever – and it just wouldn't end: in the fall a city of 20,000 burned completely to the ground.

Climate scientists and hydrologists had long warned that this could happen because of the recent rise in greenhouse gas concentrations in the atmosphere to levels unprecedented in the history of humanity and the fact that so many of the almost eight billion people on this planet live in locations prone to flood and fire. We knew the threat, but we didn't expect it to come true so soon. If you are looking for a smoking gun to confirm the acceleration of global climate change, the summer of 2018 may be it.

There are a number of immediate lessons that stand out in terms of wildfire from that summer. As we saw in Ed Struzik's *Firestorm*, discussed in chapter 3, we need to halt watershed degradation, protect and restore full watershed function, and reverse century-old forest management practices that focused on wildfire suppression and resulted in changes in forest dynamics that have led to huge fuel buildup and greater risk of catastrophic fires. But we have to also understand that it is not just the future of our forests that should concern us. And while not every year will be a fire year, there are other effects that will make themselves known over time. The new normal is that there will be no new normal.

In our view, climate warming is now a permanently lit match held over not just the forests but the entire geography of the northern hemisphere, if not the whole globe. The match is lit, and the only way to extinguish it is to restore balance to Earth System function.

If the reader is in need of further evidence of just how serious our situation has become globally, here are some numbers that may be useful. In the past 20 years, 157,000 people have died as a result of floods. Between 1995 and 2015, according to the UN Office for Disaster Risk Reduction, floods affected 2.3 billion people, droughts 1.1 billion more. Another 660 million were harmed by storms, and a further 8 million by wildfire. Recent research has demonstrated that with a mean global temperature rise of 1.5°C, human losses from flooding could rise by between 70 and 83 per cent, and direct flood damage might increase by 160 to 240 per cent. In a 2°C world the death toll is 50 per cent higher and direct economic damage doubles. Impacts at 3°C are notably harder to model, as the extent of variability becomes staggering.

Conflicted politicians, paid climate change deniers, and skeptics in the employ of covertly weaponized philanthropic organizations masquerading as legitimate democratic institutions will immediately write these claims off as "fake news." They will of course counter that the Earth has witnessed such changes in climate before. But they will refuse to acknowledge that at no time in the Earth's history was the planet occupied by nearly eight billion people trapped in infrastructure designed for an earlier and less volatile climate regime. Such protestations will not change the facts, however, nor alter the weather. Climate disruption is no longer just a threat to public health, local economies, food security and regional and national political stability.

What we are discovering is that by allowing landscapes and riparian systems to degrade over time, regions and entire nations lose the effective buffering provided by intact natural processes, leaving them exposed to the full force of increasing hydro-climatic variability. This suggests that places with limited, lost or compromised natural absorptive protection from climate impacts are far more directly vulnerable to acceleration of the global water cycle and more subject to climate disasters than places where natural systems are still in place that slow or moderate those effects.

And it is here that suddenly the damage that countries like China and we ourselves are doing to the stability of the Earth System, and in particular the effect it is having the global climate, becomes very, very relevant.

Many prominent scientists are saying that 2018 may have been a turning point in human history. During that summer we realized that because we have not taken adequate action to reduce carbon dioxide emissions, we are now facing what could very well be a planetary emergency. It bears repeating. If you were in British Columbia, Alberta, the Northwest Territories, Saskatchewan, Manitoba, northern Ontario or even parts of Atlantic Canada that summer and you wanted to know what that planetary emergency will look like in the future, all you had to do was look out the window. And Canada was not alone in this. If you were in affected parts of Asia, Europe or even Australia on August 17, 2018, and you wanted to know what climate change will mean to your future, you too had only to step outside to see what is to come. But in facing what appears to be a planetary emergency square on, there is not only opportunity to create a better world but also hope for the future. Now that we have been put on notice, what should we do?

From that summer we see that the period of being reactive and continuously debating evidence of what we thought would happen if our climate changed must now be over. Denying what anyone will soon see will unnecessarily inhibit effective action in ways that could quickly morph into anxiety and then anger and lead to widespread panic. It is the role of emerging leaders not only to undertake restoration but also to carefully direct anxiety that could turn to despair into positive action. We have to catch up and then get ahead of this problem and start solving it. Doing so will require courage. Fortunately we still have room to move.

While it is technologically possible to reduce emissions to near zero, the only way this target can be achieved would be to put the eradication of greenhouse gas emissions at the same level of importance as economic growth. This would be the criterion for any country to succeed, for Canada to succeed. As noted in the previous chapter, making restoration of Earth System stability the supreme priority is not going to be easy anywhere, especially in a country like Canada that remains indifferent to truly meaningful climate action. The more we look into the problem, however, the more we see not just the necessity for this balance but the benefits that establishing this balance could bestow on us and future generations.

So let's look at the good news. Research has revealed new and very interesting findings suggesting that extraordinary efforts to restore natural system function are not only warranted but may be our only practical and affordable way forward in the future. In making a case for a global restoration imperative, let me return, as I have above and in other writings, to Gro Harlem Brundtland's definition of sustainability: meeting the needs of the present without compromising the ability of future generations to meet theirs.

I would submit to you that our situation, as witnessed in the summer of 2018, is that in so substantially altering the fundamental composition of the global atmosphere, we have generated effects that are cascading throughout all of nature. We have literally altered Earth System function to such an extent that, if we are not careful now, the world future generations will inherit could be unlike the world any human has ever known. Fulfilling the promise of meeting the needs of our own generation without compromising the ability of future generations to meet theirs appears at the moment to be nearly impossible.

At this moment there is no guarantee we will even be able to meet our own needs in these altered conditions, let alone the next generation theirs or their children's. This means we must first – and to a very real extent – restore the world as we have known it before we can begin fulfilling the terms of Brundtland's definition of what it means to be sustainable. We literally have to go back to go forward again.

As noted earlier, the loss of hydro-climatic stability is telling us that true sustainability may be beyond our grasp if we don't do the right thing now. Many believe we will adapt and become more resilient as a society, if only out of sheer necessity. At present, however, policy change continues to move at a pace far slower than reality, while hydro-climatic disruption keeps accelerating every year. We can and must do more.

We keep talking about adaptation in service of resilience, but resilience implies protecting what we have now. To be sustainable, development in the future must be not only environmentally neutral but also both restorative and *presilient*. What this means is that policy must be responsive not just to the world as it is now but also to the very different conditions we are bringing into existence by way of the damage we are causing to the Earth's self-regulating, biodiversity-based life support functions.

Sooner or later, however, as climate disruption envelops and eventually kills more and more people, recognition of the disaster we face will lead to mass demands for immediate change. We must be ready when that happens. Interests behind the climate change denial industry and politicians with other agendas would have us believe that scaring the public with the prospect of climate change just paralyzes people. There is little foundation for this argument. One wonders what would have happened if Winston Churchill, out of fear it would frighten Londoners, had refrained from telling them the Germans were going to bomb the hell out of their city. Sooner or later they would have had to face the truth. So Churchill told all of Britain what was coming and people rose to the challenge. Why shouldn't we expect that now?

Failure to acknowledge, face and come to terms with what was clearly before everyone's eyes was in large measure what caused the Second World War, which did to history what hurricanes have done to the Florida coast. Once you have seen images of Hurricane Michael and its aftermath in the Florida panhandle, it is impossible to get them out of your mind.

How many towns and cities will have to be obliterated, how many people will have to die, before we acknowledge we have started an undeclared war against nature itself, a war we cannot win. But just as Londoners rallied to their challenge, people everywhere show us, time and again, that average citizens surprise themselves and the world with their remarkable fortitude, selflessness and sheer bravery in the most impossible situations.

Can't we give people the credit they deserve for intelligence, resilience and courage?

After the Second World War ended, the object was to make sure that such a global catastrophe would never happen again. Institutions like the UN were created to prevent conflict from becoming widespread. By way of reconstruction, a new world came into existence, a better, more stable, more peaceful world. If we are to end our war with nature before it destroys us, we must make peace with self-regulating Earth System function, reconcile the injustices of the past and make peace with one another.

It is in the will to action that our hope for the future resides and along with it a once-in-a-lifetime opportunity to create a better world for ourselves and our children and grandchildren. This is the present generation's moment. More than at any other era in human history, this is a time for heroic, committed leadership and dauntless citizenship in service of a future in which we not only survive but flourish. That leadership and that citizenship do exist, but what is missing is a sense of urgency.

We live in one of the last places on Earth where it is still possible to transcend the climate debate in time to create a better world. Canada has enough remaining natural capital to protect and restore its way back to true sustainability. If we put our mind to it, we can rehabilitate our rivers and lakes and restore our forests. We can reverse our adverse effects on phosphorus and nitrogen cycles. In so doing we can stabilize the climate, and we might even help stabilize the world.

We cannot restore lost biodiversity, but we can halt its decline and consciously direct evolution toward a richer future. We can make where we live, and the world, better.

A question I am very often asked is: What can I do? I am just one person. As it happens, there is a lot one person can do in the context of where and how they live, so there is no need to throw one's hands up in helpless despair. You can start by reducing your personal çarbon footprint. But there are also other actions that are not only helpful but can also be personally satisfying and life-enriching. Start where you live – where you have the greatest power to effect change – and work outward from there. You can make where you live – your yard, your property, your neighbourhood and your community and its surroundings – as natural and beautiful as you can make them. That is what one person can do. And if that is not enough, stop being just one person: join others with similar aspirations.

Canadians are in the midst of advancing an entirely new field of restoration hydrology, also known as applied regenerative hydrology. What this involves is using our understanding of landforms and how they are influenced by precipitation to increase soil moisture and health while at the same time providing free water storage. What we are learning is that through thoughtful ecological succession we can increase the role of natural system function in disaster risk reduction and carbon sequestration while at the same time creating healthier, safer, pleasanter and more productive places to live.

We are now shifting some investment in Canada toward restoration of upland watersheds, with the goal of viewing water infrastructure not just in terms of hard engineering but as a combination of both natural and engineering elements. In addition to improving engineered solutions, we are also learning how to use natural ecosystem function to help build more efficiently integrated water infrastructure that as much as possible operates and maintains itself. These are the kinds of changes any community can make.

This has not been an entirely happy story to have to tell. But the good news is that it doesn't have an ending yet. It is the ending I hope readers will over time help write by way of their love of place and deep commitment to the future. The rain is still comin' down, but there are still sunny days. Together we can write a different story with a much better ending.

RIVERS OF HOPE

*For Heather Read
and her Facebook friends*

Nyk nailed it.

The ultimate goal of science
is to make the invisible visible;
to make the impossible possible;
to relentlessly pursue
and iteratively confirm the truth.

But not everyone wants the truth.

Much that is invisible remains, on many planes.

Forty climate researchers
from a dozen countries
meet at a high-altitude research centre
in the Alps.

Schneefernerhaus, as it is called, is as unlikely as Shangri-La.
Plastered like a barnacle against an impossibly steep rock wall
beneath the summit of Germany's highest mountain,
it seems, like the world, to be just hanging on.

The tone there is very sober.
No one is under any illusions as to what we face.
The world order is now in free fall.
The fragile ideal of a unified world is being shattered.
Truth has again become liquid,
confined to the context of self-interest and ideology.

We know from history where this leads.

Science must do all it can, now,
to slow and moderate the effects of that free fall.
If the global scientific community fails to do so,
it is within the realm of real possibility
that climate disruption could become a planetary
    emergency
adding substantially to pressures on an already fractured,
less co-operative world.

Schneefernerhaus is perched so precariously
on such a vertical slope
that it has to be protected
by avalanche defences
which include alarmed doors
that automatically lock
when sensors deep in the snowpack pick up
heightened avalanche risk
on the walls above.

Only those who see the invisible
can achieve the impossible.

When the alarm goes off
and the doors lock,
the research lab
descends like a submarine into the roaring snows
diving into the deep subsurface
interconnectedness of our planet's climate,
where the invisible impacts
of a justifiably worried world
reverberate back and forth
like whale song
beneath the deep ocean
that is our collective consciousness.

What would you say to these submariners?
Heather asks her Facebook friends.

Tell them that we hear them
and they are not alone.
You cannot see us all behind you, but
we are here.

Tell them that regular people
look to them with deep gratitude and respect.
We have respect for the commitment
of scientists to truth.
The weight upon scientists
is a great one.
Don't give up, we need you.

We are with you, the people of the world.
Let truth prevail;
bend not to the howling winds
of political pressure and pretense.

Scientists have an uplifting effect
on the unconscious collective.
We are watching and listening
and we are ready to respond.

We want to know the most effective grassroots ways
to create change.
We know this is important work
because of the work climate scientists are doing.

What is getting in the way is the power dynamic
of a system built on privatization, commodification,
scarcity and greed.

There is more good happening
in the world
and solutions exist for the problems we face.

We continue to mobilize
a community
that will stand for our water,
our oceans, atmosphere,
forests, minerals, soil,
culture and all the beauty
of the Earth.

People all over the world are doing MORE than ever
to protect the environment;
to shift democratic process;
move away from traditional banking systems;
and hold responsible those who operate without regard
for the needs of the planet.

We are a society in transition.
People are beginning to awake
from the slumber of illusion
and together anything is possible.

Tell the scientists
that every day
thousands of women are gathering
to pioneer a new way of being in the world.

Ordinary women are devoting themselves
to staying awake to the truth
of the situation we are in.

We are growing carbon dioxide-absorbing tree plantations;
reclaiming the land for local food production;
and creating solar municipalities.

Tell the scientists
we have immense hope.

May you feel us lifting you and your work
and holding space
for the hard truths
you must shepherd.

But we, the researchers, are reciprocally buoyed,
we are lifted by you, also.
We want to be worthy of that trust,
that respect.

We are learning to use satellites,
and sensors and big computers
to see the invisible
in the air, in water, in soil,
in our past, our present,
and into the future.

The growing accuracy of data,
expanded understanding of Earth System function,
greater knowledge and emerging common urgency
are driving a revolution
in the Earth sciences.

Multi-spectral space-based remote sensing
is making the invisible visible.
Combined with careful
terrestrial ground-truthing
the impossible may soon be possible.
The Holy Grail of the hydrological sciences
is within our grasp.
Integrated flood and drought prediction and forecasting
and much, much more will soon be possible.

We may not save a single glacier
but we may find a way to restore the world.
We need people like you behind us, however;
people who believe that it remains possible
for the rupturing chrysalis of contemporary culture
to transform itself into a blue-green dragonfly
with wings white as a peace-dove.

For us, it is coming down to this:

either we will witness the greatest and most rapid
    transformation
of humanity's understanding of itself
and vision for the future of the world and the planet

or

we will have the most carefully and completely documented
collapse of a civilization in all of human history.

In part because of you –
because of your concern,
your commitment
and your voice,
for which we offer our deepest thanks

we, the scientists, are still betting with you,
not on destruction, but on transformation.

So it is together that we, all of us, continue, with gratitude,
to undertake the hard work of hope in our time.

                                     — Bob Sandford
                                     February 2018

APPENDIX

A CANADIAN NATIONAL GLACIER ACT

A minimum standards regime for the preservation of glaciers and the periglacial environment
　(an unofficial sketch)

Article 1: Subject

The following law establishes the minimum standards for the protection of glaciers and the periglacial environment with the objective of protecting them as strategic freshwater reserves for human consumption; for agriculture and as sources of watershed recharge; for the protection of biodiversity; and as a source of scientific knowledge and understanding as well as a tourist attraction.

Article 2: Definition

We understand glaciers to be all perennial stable or slow-flowing ice masses larger than 0.01 square kilometres, with or without interstitial water, formed by the recrystallization of snow, located in different ecosystems, whatever their form, dimension and state of conservation. Detrital rock material and internal and superficial water streams are all considered constituent parts of each glacier. Likewise, we understand that in the periglacial environments of high mountains, frozen ground functions as a regulator of the freshwater resource. In middle and low mountain areas and in the Arctic, it is ice-saturated ground that functions as a regulator of freshwater resources.

Article 3: Inventory

The Canadian National Glacier Inventory is hereby created, in which all glaciers and periglacial landforms that act as freshwater reserves shall be identified and recorded along with all appurtenant information that may be necessary for their adequate protection, control and monitoring as set out in Article 4.

Article 4: Registration of information

The Canadian National Glacier Inventory shall contain information about glaciers and the periglacial environment organized by watershed, location, surface area and morphological classification. This inventory shall be updated every five years, verifying changes in the total area of glaciers and the periglacial environment, their advance or retreat, and any other factors relevant to their conservation.

Article 5: Implementation of the inventory

The inventory and monitoring of the state of the glaciers and the periglacial environment shall be carried out by Canada's National Water Security Centre, in coordination with the national implementing authorities of this law. Global Affairs Canada shall participate where border zones are concerned in which international demarcation is still pending prior to inventory registration.

Article 6: Prohibited activities

All activities that could affect the natural condition or the functions listed in Article 1, that could imply the destruction or dislocation of glaciers or interfere with their advance, are prohibited, in particular the following:

1. the release, dispersion or deposition of contaminating substances or elements, chemical products or residues of any nature or volume, whether occurring on or in glaciers or anywhere in the periglacial environment;
2. the construction of works or infrastructure, except those necessary for scientific research or risk prevention;
3. mining and hydrocarbon exploration and exploitation, whether on or in glaciers or anywhere in the periglacial environment; and
4. any other works or industrial activity.

Article 7: Environmental impact evaluations

All activities planned on glaciers and in the periglacial environment that are not prohibited shall be subject to environmental impact evaluations and environmental strategic evaluations, depending on the scale of intervention, in which public citizen participation must be guaranteed as established by federal environmental impact assessment legislation before authorization and implementation is granted. The following activities are excluded from these requirements:

1. rescue activities as a consequence of emergencies;
2. scientific activities taking place on foot or on skis, with

eventual sample-taking, that do not leave waste on glaciers or in the periglacial environment; and

3. sporting activities, including trekking, mountain climbing and non-motorized sports that do not perturb the environment.

Article 8: Competent authorities

As per present law, the competent authority shall be that authority appropriate to each jurisdiction. Federally this will be a national water commission to be brought into existence under separate legislation. In the case of areas protected as national parks, the competent authority will be Parks Canada.

Article 9: Implementation authority

The implementing authority of the present law shall be the institution with the highest national environmental jurisdiction.

Article 10: Functions

The functions of the national implementing authority will be:

a) the formulation of actions conducive to the conservation and protection of glaciers and the periglacial environment in a coordinated manner with competent provincial agencies and federal ministries in their respective areas of competence;

b) contribution to the formulation of a climate change policy relative to the objective of glacier protection, that harmonizes provincial and national activities with one another and with international climate change agreements;

c) the coordination of the execution and updating of the Canadian National Glacier Inventory through the National Water Security Centre and other relevant agencies and institutions;

d) the preparation of a periodic report on the state of existing glaciers in Canada as well as projects taking place on glaciers or in their zones of influence, which reports shall be submitted to the federal cabinet;

e) the offering of advice and support to local jurisdictions in monitoring programs, controls and glacier and periglacial protections;

f) the creation of programs to promote and create incentives for research;

g) the development of campaigns to educate the public and produce environmental information conforming with the objectives of the present law; and

h) inclusion of the principal results of the Canadian National Glacier Inventory and its updates in national information sent to the United Nations Framework Convention on Climate Change.

Article 11: Infractions and sanctions

The sanctions for non-compliance with the present law and the regulations that shall be introduced, beyond other responsibilities that might apply, shall be those that are established by the jurisdiction according to its corresponding policing power and which shall not be lower than those established here. Jurisdictions that do not have a sanctions regime shall apply the following sanctions which correspond to the national jurisdiction:

a) warning;

b) fine of 100 to 100,000 times the minimum income of the average entry-level national public administration employee;

c) suspension or revocation of authorization. Suspension of activity could be from 30 days up to one year, according to the merits and circumstances of the case; and

d) definitive cessation of activities.

These sanctions shall be applicable following legislatively substantiated summary proceedings in the jurisdiction where the infraction took place; shall be regulated by the corresponding procedural administrative norms, assuring due legal process; and shall be incremented according to the nature of the infraction.

Article 12: Reincidence
In the case of reincidence, the minimum and maximum sanctions stipulated by paragraphs b) and c) of the previous article could be tripled. It shall be considered reincidence when, within a period of five years following the commission of an infraction, the party has been sanctioned for another infraction of environmental cause.

Article 13: Liability
Where the violator is a juridical person, all persons in positions of direction, administration or management shall be jointly and severally liable to the sanctions established by the present law.

Article 14: Application of the fines collected
All sums collected by the competent authority shall be directed in priority to the protection and environmental restoration of the glaciers affected in each jurisdiction.

Article 15: Transitional
Within a maximum of 60 days beginning with the coming into force of this law, the National Water Security Centre shall present to the national implementing authorities a chronogram for carrying out the inventory, which shall commence

immediately in such zones which, due to the existence of activities contemplated in Article 6, are considered a priority. In these zones the inventory stipulated in Article 3 shall be carried out within no more than 180 days.

Competent authorities shall provide all pertinent information required by the cited institution.

Activities described in Article 6 that are in progress at the moment of the coming into force of the present law, must, in a period of no more than 180 days from the promulgation hereof, submit to an environmental audit in which potential and actual environmental impacts to glaciers are identified and quantified. In the case of verification of negative impacts to glaciers or the periglacial environment contemplated in Article 2, the authorities shall order the pertinent measures so that the present law is complied with, and may order cessation or relocation of the activity; protective measures; cleaning; and restoration as appropriate.

Article 16: Coming into force

The present law comes into force 90 days from the date it is promulgated in the *Canada Gazette*.

The Periglacial in the Canadian Rockies:
Unnamed Rock Glaciers, Banff National Park

R.W. Sandford, United Nations University Institute for Water,
Environment & Health

BOOKSHELF

Alexis, André. *Fifteen Dogs*. Toronto: Coach House Books, 2014.

Barnett, Cynthia. *Rain: A Natural and Cultural History*. New York: Crown, 2015.

Carroll, Lewis. *Alice's Adventures in Wonderland*. London: Macmillan, 1928. First published 1865 by Clarendon Press. Accessed 2019-06-01 at is.gd/2qPSdF.

Carson, Rachel. *Silent Spring*. Boston: Houghton Mifflin, 1962.

Catley-Carlson, Margaret. "The International System for Water Management: Plethora, paucity, whisky and worry." *Policy Options*, July 1, 2009. Accessed 2019-06-01 at is.gd/WJCSYT.

Childs, Craig. *Atlas of a Lost World: Travels in Ice Age America*. Illustrated by Sarah Gilman. New York: Pantheon, 2018.

Diamond, Jared. *Guns, Germs, and Steel: The Fates of Human Societies*. New York: Norton, 1997.

Dillard, Annie. "The Waters of Separation." In *Pilgrim at Tinker Creek*. New York: Harper's Magazine Press, 1974.

———. "Write till You Drop." *New York Times*, May 28, 1989. Accessed 2019-06-01 at is.gd/pZ1Zmr.

———. *The Writing Life*. New York: Harper & Row, 1989, 2013.

Durrell, Lawrence. *The Dark Labyrinth*. London: Faber and Faber, 1962.

Eisenberg, Evan. *The Ecology of Eden*. New York: A.A. Knopf, 1998.

Gowdy, John, Lisi Krall and Yunzhong Chen. "The Parable of the Bees: Beyond Proximate Causes in Ecosystem Service Valuation." *Environmental Ethics* 35, no. 1 (2013): 41–55. Accessed 2019-05-30 (pdf) from is.gd/BsAufm.

InterAction Council. "Dublin Charter for One Health." 34th Annual Plenary Meeting, Dublin, Ireland, May 30–31, 2017. Accessed 2019-06-01 at is.gd/zZHJpx.

International Institute for Sustainable Development. "China's war on pollution – and what comes next." Blogpost, July 6, 2018, by Scott Vaughan. Accessed 2019-06-01 at is.gd/8uB7Vn.

Kemmis, Daniel. *This Sovereign Land: A New Vision for Governing the West.* Washington, DC: Island Press, 2001.

King, Stephen. *On Writing: A Memoir of the Craft.* New York: Scribner, 2000.

Kittredge, William. *The Nature of Generosity.* New York: Alfred A. Knopf, 2000.

Lovett, James. *O Istanbul: Poems for a Turkish Album / Ey İstanbul: Türk Albümü İçin Şiirler.* Translated into Turkish by Coşkun Yerli. Istanbul: YKY, 2001.

MacKinnon, J.B. *The Once and Future World: Nature As It Was, As It Is, As It Could Be.* Toronto: Random House Canada, 2013.

Marsh, George Perkins. *Man and Nature, Or Physical Geography as Modified by Human Action.* London: Sampson Low, Son and Marston, 1864. Accessed 2019-06-01 at is.gd/vwXm8E.

National Academies of Sciences, Engineering, and Medicine (U.S.), Committee on Extreme Weather Events and Climate Change Attribution, Board on Atmospheric Sciences and Climate, Division on Earth and Life Studies. *Attribution of Extreme Weather Events in the Context of Climate Change.* Washington, DC: National Academies Press, 2016. Accessed 2019-06-01 at is.gd/ECDMLp.

Rush, Norman. *Whites: Stories.* New York: Vintage Books, 1992.

Sagan, Carl, and Ann Druyan. *Comet.* New York and Toronto: Random House, 1985.

Sandford, Robert W. *Cold Matters: The State and Fate of Canada's Fresh Water.* Victoria and Calgary: RMB, 2012.

———. *Ecology & Wonder in the Canadian Rocky Mountain Parks World Heritage Site.* Edmonton: AU Press, 2010.

———. *Quenching the Dragon: The Canada–China Water Crisis.* Victoria and Calgary: RMB, 2018.

————. *Water and Our Way of Life*. With photographs by Steve Short. Fernie, B.C.: Rockies Network, 2003.

————. *The Weekender Effect: Hyperdevelopment in Mountain Towns*. Victoria and Calgary: RMB, 2008.

Struzik, Edward. *Firestorm: How Wildfire Will Shape Our Future*. Washington, DC: Island Press, 2017.

Taillant, Jorge Daniel. *Glaciers: The Politics of Ice*. Oxford and New York: Oxford University Press, 2015.

United Nations. *Transforming Our World*: The 2030 Agenda for Sustainable Development. General Assembly A/RES/70/1, September 25, 2015. Accessed 2019-06-01 at is.gd/aE8s1n.

Vonnegut, Kurt. *Cat's Cradle*. New York: Holt, Rinehart and Winston, 1963.

Watts, Robert J. "The Basarwa Problem: An Examination of the Incorporation of a Fourth-World People into the Nation-State of Botswana." MA thesis submitted to the Department of Sociology and Anthropology, Simon Fraser University, 1984. Accessed 2019-05-30 (pdf typescript) at is.gd/XKM0ii.

Williams, Terry Tempest. "Home Work." In *Red: Passion and Patience in the Desert*. New York: Pantheon Books, 2002.

————. *Refuge: An Unnatural History of Family and Place*. 10th anniversary edition with a new preface and afterword. New York: Vintage, 2001, 2018. First published 1991 by Pantheon.

————. *When Women Were Birds: Fifty-four Variations on Voice*. New York: Farrar, Straus and Giroux, 2012.

————. "Winter Solstice at the Moab Slough." In *An Unspoken Hunger: Stories from the Field*. New York: Vintage Books, 1994.

Winchester, Simon. *The Man Who Loved China: Joseph Needham and the Making of a Masterpiece*. New York: HarperCollins, 2008.

Workman, James G. *Heart of Dryness: How the Last Bushmen Can Help Us Endure the Coming Age of Permanent Drought*. New York: Walker, 2009.

Worster, Donald. *Rivers of Empire: Water, Aridity, and the Growth of the American West*. New York: Pantheon Books, 1985.

Wright, Ronald. *A Short History of Progress*. Toronto: Anansi, 2004.

Zabor, Rafi. "The Turn: Inside the secret dervish orders of Istanbul." *Harper's Magazine* 398, no. 1849 (June 2004): 49–63. Accessed 2019-05-30 (pdf page images) from EBSCOhost MasterFile Premier, accession no. 13162658, via Calgary Public Library proxy server (library card required).

BOOKS BY
ROBERT WILLIAM SANDFORD

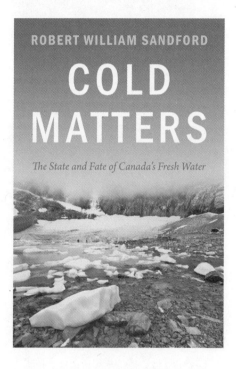

COLD MATTERS
The State and Fate of Canada's Fresh Water

9781927330197, softcover

This timely book gives the concerned reader an opportunity to take part in the conversation about our global environment in a way that transcends traditional scientific journals, textbooks, public talks or newspaper articles that are so often ignored or forgotten. In the end, *Cold Matters* will change the way you think about ice and snow.

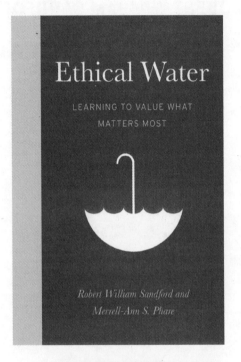

ETHICAL WATER
Learning to Value What Matters Most

9781926855707, hardcover

This ground-breaking and approachable work, by two of Canada's most authoritative experts on water issues, redefines our relationship with fresh water and outlines the steps we as a society will have to take if we wish to ensure the sustainability of our water supply for future generations.

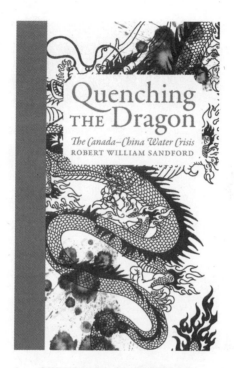

QUENCHING THE DRAGON.
The Canada–China Water Crisis

9781771602938, hardcover

Part environmental manifesto, part travelogue and part diplomatic odyssey, *Quenching the Dragon* arms readers with vital new perspectives on global hydrology and sustainability in the context of how former and current world leaders frame the most pressing environmental issues of our time.

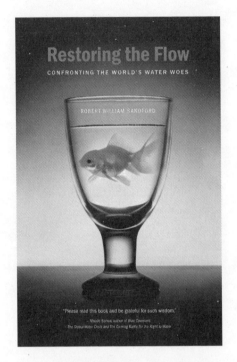

RESTORING THE FLOW
Confronting the World's Water Woes

9781897522523, softcover

Passionately conceived, clearly written and citing concrete examples from all over the world, *Restoring the Flow* is an approachable yet authoritative source, one of the many implements concerned citizens, government officials, business people and policy-makers can use and reuse in understanding and addressing this ever-growing global crisis.

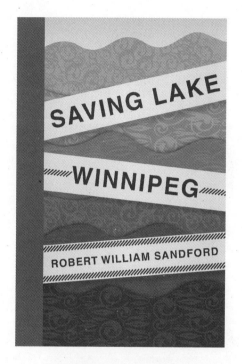

SAVING LAKE WINNIPEG

9781927330869, hardcover

The toxins produced by algae blooms in Lake Winnipeg each summer are a growing threat to human health, agriculture and the general economy in both Canada and the US. Robert Sandford's third RMB Manifesto speaks on behalf of the Central Great Plains to convince government, industry and society that drastic change is needed if we are to avoid similar troubles in other bodies of water throughout North America.

STORM WARNING

Water and Climate Security in a Changing World

9781771601450, softcover

This highly considered, accessible and readable book explains how changes in the water cycle have already begun to affect how we think about and value water security and climate stability and what we can do to ensure a sustainable future for our children and grandchildren.